NATURE'S MIRROR

NATURE'S MIRROR

How Taxidermists Shaped America's Natural History Museums
and Saved Endangered Species

MARY ANNE ANDREI

THE UNIVERSITY OF CHICAGO PRESS
CHICAGO AND LONDON

The University of Chicago Press, Chicago 60637
The University of Chicago Press, Ltd., London
© 2020 by The University of Chicago
Published 2020
Printed in the United States of America

29 28 27 26 25 24 23 22 21 20   1 2 3 4 5

ISBN-13: 978-0-226-73031-8 (cloth)
ISBN-13: 978-0-226-73045-5 (e-book)
DOI: https://doi.org/10.7208/chicago/9780226730455.001.0001

Publication of this book has been aided by a grant from the Neil Harris Endowment Fund, which honors the innovative Scholarship of Neil Harris, the Preston and Sterling Morton Professor Emeritus of History at the University of Chicago. The fund is supported by contributions from the students, colleagues, and friends of Neil Harris.

Library of Congress Cataloging-in-Publication Data

Names: Andrei, Mary Anne, author.
Title: Nature's mirror : how taxidermists shaped America's natural history
   museums and saved endangered species / Mary Anne Andrei.
Description: Chicago : University of Chicago Press, 2020. | Includes
   bibliographical references and index.
Identifiers: LCCN 2020003829 | ISBN 9780226730318 (cloth) |
   ISBN 9780226730455 (ebook)
Subjects: LCSH: Taxidermy—United States—History. | Natural history museums—
   United States—History. | Zoological specimens—Collection and preservation—
   United States—History.
Classification: LCC QL63 .A547 2020 | DDC 590.75/2—dc23
LC record available at https://lccn.loc.gov/2020003829

♾ This paper meets the requirements of ANSI/NISO z39.48-1992
(Permanence of Paper).

To hold, as 'twere, the
mirror up to nature; to show virtue her own feature,
scorn her own image, and the very age and body of
the time his form and pressure.
—Hamlet to the players (*Hamlet*, act 3, scene 2)

CONTENTS

Introduction

> Our museums have an enviable reputation for the manner in which they hold the mirror up to Nature, and yet I feel that the [Ward's Natural Science] Establishment may justly claim a large share of the credit for this.
> —Frederic A. Lucas[1]

A throng of more than two thousand crowded into the rotunda of Theodore Roosevelt Memorial Hall at the American Museum of Natural History (AMNH). Flashbulbs popped, and reporters from all the New York dailies pushed through the room, scribbling furiously in their notepads. It was May 19, 1936—a date chosen by the museum with care. Had he lived to see it, that day would have been the seventy-second birthday of Carl Ethan Akeley, the legendary taxidermist and conservationist who, twenty-five years earlier, had first conceived of a great hall of African wildlife. Now, in his honor, the AMNH was, at last, realizing his dream. Daniel E. Pomeroy, a member of the board of trustees who had accompanied Akeley on his final, fatal trip to the Belgian Congo ten years before, cut the white ribbon at the threshold, and the crowd pressed forward into the Akeley Hall of African Mammals.

Inside, the lights were dimmed, one reporter wrote, "like dusk in the jungle,"[2] and a recording played the faint beat of distant drums. Rising toward the ceiling in the center of the room stood an enormous group of elephants—eight in all, the largest display of pachyderms ever attempted. The fourteen wall cases ringing the room glowed from within, combining to form a "colorful panorama of swamps, mountains and jungles and deserts, animal life and settings unusual in their artistic beauty, dramatic realism and scientific accuracy."[3] Russell Owen, an adventure journalist for the *New York Times* who had won a Pulitzer Prize by shadowing Amundsen and

Byrd on their polar expeditions, was especially impressed by the ambition and authenticity of Akeley's taxidermy. He singled out the group of gorillas as "the most striking animal assembly ever put together."[4] In tribute to the taxidermist, the five gorillas were posed atop Mount Karisimbi, the place where Akeley died of a sudden fever and was buried in November 1926, but the idyllic scene was also a monument to his lifelong ambition to portray a seamlessly realistic setting. "One can almost hear the drip of water from recent rains," Owen wrote.

Fifty years earlier, such realism in museum taxidermy and public display could only have been dreamed of. At that time, Akeley was an ambitious young man, still in his early twenties, under the employ of Henry A. Ward, proprietor of Ward's Natural Science Establishment in Rochester, New York. Akeley had first come to Ward's because, he later remembered, "Professor Ward was the greatest authority on taxidermy of his day."[5] Already his workshop had produced William T. Hornaday and Frederic A. Lucas, then the respective heads of taxidermy and osteology at the U.S. National Museum (today the National Museum of Natural History), and Charles H. Townsend, head of taxidermy at the Philadelphia Academy of Natural Sciences. These men, along with Frederic S. Webster, had formed the Society of American Taxidermists in 1880, and Webster, the society's first president, now ran Ward's shop.

Akeley had grown up on a farm near Clarendon, New York, less than twenty-five miles from Rochester. In the fall of 1883, after helping his father complete the harvest, the nineteen-year-old walked three miles to the train station, then rode into Rochester with no clear idea where he was going. "I walked all over town," Akeley recalled, "and the more I walked the lower and lower my courage sank." Finally, he found the entrance to the grounds of Ward's and entered through an archway formed by the jaws of a sperm whale. "An apprentice approaching the studio of a Rembrandt or a Van Dyke could not have been more in awe than I was," Akeley later wrote. But after pacing in front of Professor Ward's door for several long moments, he finally mustered the courage to ring the bell and was received into Ward's study, where he found the professor.

Ward had thinning gray hair, and his closely trimmed beard had gone grizzled. He was already at work, going over the morning's correspondence, while still finishing the last of his breakfast. His earnest blue eyes examined Akeley carefully through a pair of gold-rimmed glasses. "What do you want?" Ward snapped. By now, Akeley's youthful confidence had failed completely. He silently handed Ward his card: "Carl E. Akeley—Artistic Taxidermy in All Its Branches." As was his custom, Ward asked a few pointed questions, and then, when he was satisfied of the young man's seriousness, offered him a job on the spot.

Thus, Akeley became the last—and ultimately the most influential— member of a uniquely important cohort of taxidermists who trained together at Ward's in the 1870s and 1880s. Their new methods would revolutionize the world of museum display and, as a result, would permanently reshape the public's understanding of the natural world. Their work would lead to the creation of America's major zoos and lay the foundations of the modern wildlife conservation movement. First, however, this small group of colleagues and competitors had to completely remake the craft of taxidermy— from the hackwork of back-alley curio shops into a skilled discipline respected equally for its artistic excellence and its scientific accuracy.

The task was formidable. Akeley remembered that at the beginning of his career, the profession had "very little science and no art at all."[6] To underscore the point, he often recounted his half-joking belief that taxidermy had emerged when "some old-fashioned closet naturalist who knew animals only as dried skins" took them to the corner upholstery shop. "Here is the skin of an animal," he imagined the naturalist telling the proprietor. "Stuff this thing and make it look like a live animal."[7] Akeley's point, although characteristically hyperbolic, was simple: the average nineteenth-century taxidermist was no better prepared than the upholsterer to mount lifelike, scientifically accurate specimens. Most had minimal training, little understanding of anatomy, and no field knowledge of the animals they were trying to portray.

But the early history of taxidermy was neither so simple nor so brief as Akeley made it out to be—as Frederic A. Lucas well knew. Lucas was not only the director of the AMNH when Akeley's work on its African hall was undertaken; he was also Akeley's predecessor by some fifteen years in the old taxidermy shop at Ward's. "It was probably during his stay at Ward's," Lucas wrote, "that Akeley reached the conclusion that the taxidermist had evolved from the upholsterer (as a matter of fact I have been asked 'Who upholstered that specimen?')."[8] Lucas, too, had learned Old World methods at the elbow of Jules F. D. Bailly, Isidore Prevotel, and other European taxidermists at Ward's—and, like Akeley, Lucas had grown frustrated by their limitations and had pioneered new methods. But he had also developed respect for the accomplishments of his forebears.

"Some of our younger museum men, installing their striking habitat groups," Lucas wrote, "do not realize that these were foreshadowed a century or more ago nor give the earlier men credit for what they did in the face of many obstacles. What would the present generation accomplish if it had to work in rooms that relied upon fireplaces for heat and candles for light?"[9]

Voyages of scientific exploration launched from all parts of Europe in the late seventeenth century marked the beginning of a new era of systematic biological research, in which scientists were able to amass large collections of animal specimens for comparative study. Unfortunately, early naturalists opening crates and barrels after long expeditions often found little more than brittle skins devoid of their once brilliant feathers or rich fur. Even those specimens that did survive transport typically decayed or were destroyed by insects soon after they were stuffed and put on display in museum cabinets.

In a quest to extend the useful lives of study skins, naturalists tested many crude preservation techniques. As early as 1628, the English collector James Petiver instructed anyone intending to send him small birds to stuff them with flax or hemp fibers mixed with pine pitch or tar. Others used brandy as a fluid preservative; salt, alum, or lime to absorb moisture; and pungent spices or strong-smelling camphor to discourage insects. Though these preservatives protected against infestation and decay, they caused serious damage to skins, feathers, and fur, often rendering them useless for scientific study.[10]

In the eighteenth century, after years of struggling with inadequate methods at the Jardin du Roi, French naturalist René-Antoine Ferchault de Réaumur set out to develop a new standard for preparing scientific specimens. Réaumur studied all available literature on techniques of preservation and taxidermy and concluded that insect infestation created the "main impediment" to the development of ornithology, if not to all branches of natural history, as scientists depended on museum collections for description and classification. In 1748, in the *Philosophical Transactions of the Royal Society*, Réaumur published a brief manual on his preferred preservation methods, and he put the effectiveness of his techniques on full display in 1755, when he mounted a baby elephant, perhaps the first specimen of its kind in all of Europe.[11]

A few years later, Étienne-François Turgot, building on Réaumur's work, issued an even more detailed pamphlet, describing the proper methods for collecting, preserving, and mounting natural history specimens, with a special emphasis on skinning and packing birds for transport. Neither Réaumur nor Turgot, however, answered the question of how best to conserve natural history specimens once they were finally housed in museum collections. In the early 1770s, Jean-Baptiste Bécoeur, a French apothecary and naturalist, found a way. He discovered that arsenical soap was a successful preservative against insects—particularly the skin-eating *Dermestes* beetles that laid waste to countless collections. His arsenical soap, he contended, would greatly benefit museum collections, because "not only is it applicable to dried animals but also furs, woolens, anatomical pieces; in a word, to anything subject to being consumed by insects." Although several naturalists at this time published other methods for preservation, Bécoeur's arsenical soap eventually became the poison of choice, popularized at the beginning of the nineteenth century by the French taxidermist Louis Dufresne of the Muséum National d'Histoire Naturelle when he published a treatise on taxidermy in the *Nouveau dictionnaire d'histoire naturelle* (1803–4).[12]

Across the Atlantic, Charles Willson Peale, arguably America's first museum taxidermist, pioneered the use of arsenical soap in the New World by using it to preserve both bird and mammal taxidermy mounts exhibited in his Philadelphia museum, established in 1784. His specimens achieved unusual permanence, and he arranged them in striking, lifelike poses that delighted his ever-growing public—though Peale was so beset by "imprudent visitors" who couldn't believe his mounts weren't alive that he had to post signs throughout his museum warning, "Do not touch the birds for they are covered with arsenic Poisoning." As a more permanent solution to this problem, Peale began enclosing delicate specimens in glass, and finally, in 1802, moved his entire museum—in a procession led by workers shouldering the American bison—to the second floor of Independence Hall, where all could be safely encased.[13]

Peale built evenly rowed, floor-to-ceiling shelves in the museum's Long Room and arranged his bird specimens according to "order and genus with numbers to correspond with the number on each bird, and the classical name then followed and the name in French and English." He also incorporated a new innovation of display:

> It is not the practice, it is said, in Europe to paint skyys & Landscapes in the cases of birds and other animals, and it may have a neat and clean appearance to line them only with white paper, but on the other hand it is not only pleasing to see a sketch of a Landscape, in some instances the habits of the animal may be also given; by shewing the nest, hollow, cave or the particular view from whence the[y] came. There are examples of this kind in the Museum.[14]

Some of his most ambitious raptor mounts, such as one of his bald eagles, depicted his specimens on their nests, fresh-killed prey yet in their talons,

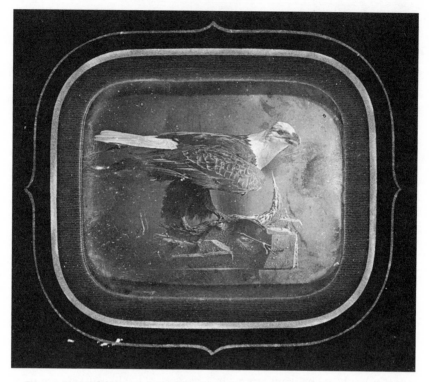

Fig. 0.1. A period daguerreotype of Charles Willson Peale's bald eagle group from his Independence Hall museum. Peale depicts the bald eagle preparing to eat a small songbird it holds in its talons. (Library of Congress)

and their beaks open as if to defend against an intruder. Frederic A. Lucas later hailed Peale as a "universal genius," whose exhibits were so ahead of their time that even "had Peale lived a hundred years later he would have been a leader in museum methods."[15]

Though Briton William Bullock probably never visited Peale's museum, he was well aware of the institution and mimicked many of Peale's methods of preservation and display. Bullock's London Museum was home to more than fifteen thousand works of art, cultural objects, specimens of natural history, and curiosities collected from around the world. The high-domed skylight in the grand room flooded the exhibit space with natural light. Below—preserved with a powder of arsenic, burnt alum, tanners' bark, camphor, and musk—stood a group of mounted African species, including an elephant, a rhinoceros, a Cape buffalo, a zebra, a lioness, and two ostriches, all fenced into a large but tightly packed enclosure.[16]

In 1812, Bullock moved the museum to an even grander building, which he dubbed Egyptian Hall for its elaborate façade of sphinxes, winged suns, and statues of Isis and Osiris over the entrance. Inside he unveiled the ambitious "Pantherion," a new hall of natural history, which was designed, he boasted, "on a plan entirely novel, intended to display the whole of the known Quadrupeds, in a manner that will convey a more perfect idea of their haunts and mode of life than has hitherto been done, keeping them at the same time in their classic arrangement, and preserving them from the injury of dust and air."

Visitors entered through a basaltic cavern based on Fingal's Cave, off the coast of Scotland, and emerged through a hut into a "Tropical Forest," including an orange tree loaded with "sixty species of the genus Simia; consisting of Apes, Baboons, and Monkeys." Beyond lay large rocks forming the dens of the "feline tribe," including the celebrated tableau of a Bengal tiger struggling to free itself from the deadly embrace of a boa constrictor; the tiger's head turned to face the visitor—its mouth open wide, tongue lolling—as the snake anchored its suffocating hold with fangs sunk deep into the tiger's neck. At the center of the room stood the African species, including a recently acquired giraffe ("the finest ever brought to Europe, and . . . in the most perfect preservation").[17]

All of the hall's specimens were exhibited "as ranging in their native wilds and forests; whilst exact models, both in figure and colour, of the rarest and most luxuriant plants from every clime give all the appearance of reality; the whole being assisted with a panoramic effect of distance and appropriate scenery affording a beautiful illustration of the luxuriance of a torrid clime." When Lucas read this description in 1921, he noted, "This seems very much like a description of some recent habitat group. Even today it is a courageous curator well provided with funds that would attempt to show the great mammals of Africa; but here is an exhibit, made by a private individual a century ago, years before Livingstone had even touched the edge of Darkest Africa, that included the largest known mammals."[18]

But even the considerable advances and lurid appeal of Bullock's artistic and blood-drenched exhibits were soon superseded by the scientific productions of Paris's Maison Verreaux. Founded by Jacques Philippe Verreaux on the Place des Vosges in 1803, the establishment quickly became the premier supplier of natural history specimens for museums worldwide (particularly the Muséum National d'Histoire Naturelle), funding scientific collecting expeditions and offering for sale to museums "thousands of species of birds, eggs and nests, as well as mammals, shells, reptiles, amphibians, and insects." Verreaux's dedication to the advancement of science through the large-scale collection and preservation of both scientific specimens and taxi-dermied museum mounts established a business model that other natural-ists would soon follow, but the greatest steps forward in the field of taxi-dermy were made by his two sons.

In 1818, eleven-year-old Jules Verreaux accompanied his uncle, Pierre-Antoine Delalande, naturalist-explorer of the Muséum National d'Histoire Naturelle, on a collecting expedition to the Cape of Good Hope. The suc-cessful two-year expedition inspired Jules to return to Africa in 1825 to pur-sue his scientific interests in ornithology.[19] Jules also worked as a curator at the South African Museum under Sir Andrew Smith—who played host to Charles Darwin, then the naturalist aboard HMS *Beagle*, in 1836. On Verreaux's return voyage to France in 1838, his ship ran aground in the Bay of Biscay. Although Jules was able to escape the wreck and swim ashore, his collection of specimens and his field notes were lost. Undeterred, he embarked on a five-year collecting expedition to Tasmania and Australia in 1842—and returned this time with his specimens and notes intact.

Jules worked for the rest of his life at the Muséum National as a collector-taxidermist while he ran the Maison Verreaux with his brother Édouard, who took control of the business during Jules's long absences. Édouard, too, was an accomplished sculptor and taxidermist, whose masterpiece, "Arab Courier Attacked by Lions," was unveiled at the Exhibition Universelle, held in Paris in 1867. It was a sensational piece, depicting a mail courier and his dromedary overtaken by Barbary lions. In the exhibit, the rider has killed one of the lions, but has dropped his rifle and is now, with only a dagger as his protection, locked in a mortal struggle with the second lion. Though Charles Wyville Thomson, chair of natural history at the Queen's College in Belfast, reported to the Crown that the Maison Verreaux showed only "a few samples of stuffing which are scarcely worthy of his world-wide reputation," the French judges disagreed—and awarded Verreaux the gold medal. The American Museum purchased "Arab Courier," along with another lone male lion also mounted by Verreaux, as the first and second items in its collection and placed them on exhibition at the Arsenal Building in Central Park.

"This group may have been theatrical and 'bloody,'" Lucas conceded, "but, as a piece of taxidermy, it was the most ambitious attempt of its day. Moreover it was an attempt to show life and action and an effort to arrest the attention and arouse the interest of the spectator, a most important point in museum exhibits. If you cannot interest the visitor you cannot instruct him." The Verreaux brothers established a standard and style for exhibition in America, but more than that, Lucas noted, "the *Maison Verreaux* suggested

to Professor Henry A. Ward the possibility of establishing a similar institution in the United States."[20]

In the waning days of 1859, as preservation ceased to be a concern and
taxidermy approached new heights, Charles Darwin at last published the
results of his voyage aboard the *Beagle* in his watershed volume *On the
Origin of Species*. Darwin's new theory of natural selection was predicated
on the idea that individuals of a species are unique and that variation from
one individual to another is the locus of evolution. Single male and female
specimens of each species, gathered into museums as if they were latter-day
Noah's arks, would no longer do. Taxonomists needed a fuller understanding
of all types of variation in order to assess whether individuals were merely
diverse members of the same species or truly distinct. As scientists further
came to understand subspecies, they recognized that geography played the
dominant role in variation. To accurately differentiate between species and
subspecies, field naturalists not only had to collect multiple specimens, but
had to obtain them from several local populations to assess the range of individual variation.[21]

This new understanding of speciation ushered in an era of broad-ranging
and intensive biological surveys in America—and brought specimens pouring into collection storage rooms. By the 1880s, America's museums were
strapped for space. To accommodate these vast new collections of skins,
study skeletons, and organs preserved in fluid, curators sought out more
space-efficient ways to store specimens. The solution was simple—but radical. American natural history museums began to divide the specimens kept
in backroom collections for scientific study from the specimens placed on
public exhibition to educate a general audience. Thus, display specimens
were no longer prepared for scientific research, but rather as dynamic repre-
sentations of living animals, intended to educate the public and to capture
their attention—and dollars—for the museums.[22]

At the same time, because circuses and newly established zoos were
bringing exotic species to Americans, the public expected something better
than the old, rough methods of taxidermy. Museum visitors were growing
too sophisticated for specimens stuffed—literally stuffed—with as much
cotton, hemp, straw, or excelsior as the skin could hold and arrayed in endless
rows of unrelated organisms. They recognized these collections for what they
were: rude imitations that bore little resemblance to, and no explanation of,
their living counterparts. A niche emerged for a new kind of taxidermy—one
that emphasized accurate anatomy and aesthetic design that could compete

Fig. o.2. Ward's Natural Science Establishment, as drawn by Frederic A. Lucas for *Ward's Natural Science Bulletin*, 1883. (Courtesy of the Department of Rare Books, Special Collections and Preservation, University of Rochester River Campus Libraries)

for the attention of a public newly acquainted with the movements and expressions of living exotic species. Undertaking such work, however, required a team of skilled and highly trained workers, often too expensive for museums to employ; this demand, in turn, gave rise to private suppliers that specialized in scientific specimens and mounts.[23]

Of all of these dealers, none was larger or more successful than Ward's Natural Science Establishment in Rochester, New York. Ward's, as it was commonly known, achieved fame by mounting bison for Buffalo Bill and the African elephant Jumbo and other circus animals for P. T. Barnum. Ward's taxidermists mounted major installations for the Centennial Exposition in Philadelphia and the 1882 Milwaukee Industrial Exposition. In 1893, Ward's brought the largest of all displays to the Chicago World's Fair. It took scores of workers months to prepare, and when packed onto the train in Rochester, the natural history specimens alone occupied thirty rail cars. In his study of nineteenth-century museum suppliers, Mark V. Barrow Jr. contends that "more than any other single institution, Ward's . . . provided the specimens that helped fuel the American museum movement."[24]

But the reputation of Ward's rested on more than its specimens. Proprietor Henry A. Ward was a prototypic nineteenth-century American entrepreneur. Combining scientific knowledge, marketing savvy, tenacity, and no little amount of ambition, Ward made a lifelong habit of flouting expectations and challenging convention. When his business, with all of his collections, burned to the ground in 1869, Ward simply rebuilt, this time with the dogged determination to make the new business bigger, better. He would

collect more and rarer specimens than any other dealer, take on larger con-
tracts, mount more intricate and expensive exhibits, all to showcase the
might of his establishment. Many years later, one of Ward's most prominent protégés, William T. Hornaday, would reflect that Ward "did more to
inspire, to build up, and to fill up American museums than any other ten
men of his time—or since his time."[25]

Though Ward's enjoyed decades as the preeminent natural history supplier in the New World (and continues to supply educational items for science classrooms), its greatest legacy remains the role it played as an educational institution for the most influential taxidermists, museum builders,
and early conservationists in American history. The men who worked at
Ward's in the 1870s and 1880s went on to become the first chief taxidermists at nearly all of America's leading metropolitan natural history museums, including the Smithsonian Institution's U.S. National Museum, from
1882 to 1920; the Milwaukee Public Museum, from 1887 to 1895; the Field
Museum in Chicago, from 1893 to 1909; and the Carnegie Museum in Pittsburgh, from 1897 to 1939. At New York's American Museum, three different Ward's trainees ran the taxidermy department from 1900 to 1948, and
another served as director from 1911 to 1929. Several other Ward's taxidermists went on to directorships at the Milwaukee Public Museum, the Bronx
Zoo, the New York Aquarium, the Brooklyn Institute of Arts and Sciences,
and the National Zoo.

More than mere taxidermists, these men were, in the words of Frederic A. Lucas, "all-round" naturalists—experienced field collectors with a
working knowledge of anatomy, osteology, and taxonomy. They studied
wildlife in the field, recording the habitats and behavior of their subjects,
taking precise measurements of their specimens, and preparing comparative
anatomical data. Such care not only led to new methods in taxidermy, but
also provided data for scientists and contributed directly to growing public
awareness of the devastating effects of careless human interaction with the
natural world.

Most nineteenth-century naturalists' knowledge of and passion for the
natural world began with hunting, and many never lost their enthusiasm for
the chase. According to historian John F. Reiger, the post–Civil War era—a
period during which the natural world was increasingly threatened by "rapid
industrialization and urbanization"—saw the American wildlife conservation movement begin in earnest. In particular, recreational "sportsmen"
of the upper classes argued that wildlife needed protection from the lower-
class "pot" (meat) and "market" (commercial) hunters who wantonly pursued wildlife for personal gain, without an appreciation for the aesthetics of

the sport, nor adherence to game laws or a "sportsman's code of ethics."[26] These gentleman hunters believed that the "true sportsman" had an agreement with his quarry, which included restricting hunting during breeding season, allowing game a "fair chance" of escape, and bag limits to ensure the survival of game species and future hunts.[27]

As the Ward's cohort explored the natural world, hunting and collecting museum specimens, their understanding of the role of the hunter in species conservation evolved beyond the "sportsman's code of ethics." They regarded themselves as museum men, separate and apart from sportsmen, who hunted in the service of science. As a result of their field work, the museum men had firsthand knowledge of threatened species and their diminishing numbers—and many felt compelled to educate the public about the destruction of wildlife by overhunting.

In *Our Vanishing Wild Life*, Hornaday beseeched, "I am now going to ask both the true sportsman and the people who do not kill wild things, awake, and do their plain duty in protecting and preserving the game and other wild life which belongs partly to us, but chiefly to those who come after us . . . before it is too late." Hornaday described the "Army of Destruction" as a "motley array" that included "regular sportsmen beside ordinary gunners, game-hogs and meat hunters."[28] He defined the "true sportsman" as "a man who protects game, stops shooting when he has 'enough'—not to take up to the legal bag-limit because he can, and whenever a species is threatened with extinction, he conscientiously refrains from shooting it."[29]

Hornaday's views of "gunners who kill to the limit" were undeniably bound to his tendency to blame environmental crises on anyone who wasn't white, male, Christian, and wealthy. He decried the "Shylock spirit which prompts men to kill all that 'the law allows.'"[30] He denounced "vain and hard-hearted women" who wore feathered hats.[31] He wrote that Italian immigrants were "pouring into America in a steady stream" and warned that "the Italian laborer is a human mongoose. Give him the power to act, and he will quickly exterminate every wild thing that wears feathers or hair."[32] He proposed a code of ethics that included the proviso: "An Indian has no more right to kill wild game . . . than any white man in the same locality."[33] He dedicated an entire chapter of *Our Vanishing Wild Life* to the "destruction of song birds by southern Negroes and poor whites."[34]

Hornaday recognized that his accusations were inflammatory, even by the standards of a century ago. "Whenever the people of a particular race make a specialty of some particular type of wrong-doing, anyone who pointedly rebukes the faulty members of that race is immediately accused of 'race prejudice,'" he wrote. "I shall strenuously deny the charge. . . . Zoologically, however, I am strongly prejudiced against the people of any race, creed, club, state or nation who make a specialty of any particularly offensive type of bird or wild animal slaughter; and I do not care who knows it." And yet, Hornaday repeatedly invoked Rudyard Kipling's imperialist poem in calling it "the white man's burden" to lead the way on conservation. "The protection of wild life is now a gentleman's proposition," he wrote.[35]

Hornaday's views highlight the complex roots of many institutions of culture and education, whose founders often regarded it as their mission to provide a bulwark against the deleterious effects of immigration, integration, and equality among genders. Many aristocratic patrons and museum administrators saw themselves as the saviors of wildlife, not from the ravages of large-scale slaughter by white hide hunters, feather collectors, and commercial fishermen, but from the cook-pots of what they considered lower classes and inferior races. Many of Hornaday's assertions are undeniably racist, but his rhetoric also effectively shamed the urban ruling classes into dedicating some of their wealth to wildlife conservation efforts. "There are many men so selfish, so ignorant and mean of soul that even out of well-filled purses they would not give ten dollars to save the whole bird fauna of North America from annihilation," he wrote. "As soon as you find one, waste no time upon him. Get out of his neighborhood as quickly as you can, and look for help among real MEN."[36]

Hornaday attributed his own lack of politesse to the urgency of the cause. "Things are not as they were thirty years ago," he wrote in 1912. "To allow a great and valuable wild fauna to be destroyed and wasted is a crime, against both the present and the future. If we mean to be good citizens, we cannot shirk the duty to conserve. We are trustees of the inheritance of future generations, and we have no right to squander that inheritance. If we fail our plain duty, the scorn of future generations surely will be our portion."[37] As Hornaday saw it, natural history museums and zoos not only had a shared mission to collect, catalogue, and describe every species; they also had the pressing obligation to enlighten an uneducated general public and to win them over to the cause of conservation.

For all of his obvious shortcomings, Hornaday made major contributions to the early wildlife conservation movement. He collected and mounted American bison for the National Museum, wrote *Extermination of the American Bison* (considered by many to be the first book of the American wildlife conservation movement), initiated the first captive breeding program at the National Zoo, and teamed up with President Theodore Roosevelt to found the American Bison Society, which oversaw the reintroduction of wild bison to Yellowstone National Park. And publication of his book *Our*

Vanishing Wild Life signaled the beginning of a new era of conservation. In 2010, historian Douglas Brinkley wrote, "What Upton Sinclair's *The Jungle* had been for meatpacking reform, *Our Vanishing Wild Life* was for championing disappearing creatures like prairie chickens, whooping cranes, and roseate spoonbills."[38]

Hornaday was hardly alone in his efforts. The taxidermists from Ward's were a cohort of which Hornaday was more representative than unique. Frederic A. Lucas, Hornaday's first instructor at Ward's, collected the remains of the extinct great auk to construct re-creations for the National Museum, wrote *Animals of the Past* (the first popular book to educate the public about extinction), and used the example of the auk to lobby for legislation to protect a range of endangered species from whales to songbirds. Charles H. Townsend pioneered the breeding of northern elephant seals in captivity at the New York Aquarium, lobbied for pollution controls in New York City's waterways, and was a vocal crusader for conservation as a member, along with Lucas, of the Department of Commerce and Labor's Fur Seal Advisory Board. Frederic S. Webster collected brown pelicans and American flamingos in Florida for the Carnegie Museum and played a role in the establishment of the first National Wildlife Refuge, which protected their breeding grounds. Carl E. Akeley collected mountain gorillas in the Belgian Congo for his exhibit at the American Museum and recorded behavioral data now considered to be the foundation of modern primatology. He successfully lobbied for the creation of the Parc National Albert, the first wildlife preserve on the African continent, credited with saving the species.

The active participation of these naturalist-taxidermists—not only through the educational exhibits they created, but also through the field work, popular writing, and lobbying they undertook—established a vital leadership role in the early conservation movement for American museums, zoos, and aquaria, a role that continues to this day. Through their individual research expeditions and their collective efforts to create an ethic of global environmentalism, the men of Ward's, more than any other single group, created our popular understanding of the animal world. For generations of museum visitors, they turned the glass of an exhibition case into a window on nature—and also a mirror in which to reflect on our responsibility for its conservation.

"A Gathering Place for Amateur Naturalists": Ward's and the Birth of the Habitat Group

Ward's was a gathering place for amateur naturalists trying to find themselves. Museums were few and they were glad to get a chance anywhere. They worked at taxidermy, osteology, making plaster casts of important fossils, identifying minerals and shells and even helping with the rough work that had to be done. Most of them later won recognition as naturalists, explorers, college professors, museum directors and authors.
—Charles H. Townsend[1]

In November 1873, nineteen-year-old William T. Hornaday stepped from his train onto the platform at Auburn Station in Rochester, New York. Hornaday hadn't been outside the state of Iowa since he was barely a year old, but now he had traveled cross-country, alone, to come here. He made his way from the station across the bridge spanning the Genesee River and down College Avenue toward a tight cluster of white clapboard buildings. Though "almost in the shadow of the main building" of the University of Rochester, Hornaday later remembered, the compound was shielded from the prying eyes of curious students by large overshading elms and spreading maples, and a protective fence bordered the property on all sides. The only entrance was through an enormous Gothic archway formed by the tallowy lower jaws of a whale. There was no mistaking it; this had to be Ward's Natural Science Establishment.[2]

Only months before, Charles E. Bessey, Hornaday's professor at Iowa State College, had read aloud to him from an article in the *American Naturalist*. The author insisted, "Every scientific man should visit Professor Ward's place at Rochester, New York, and see the bee-hive of industry he has built up around." Hornaday was electrified by what Bessey read. More than thirty years later, he still remembered the description of Ward's as

Fig. 1.1. William T. Hornaday trained at Ward's Natural Science Establishment from 1874 to 1881. (Courtesy of the Department of Rare Books, Special Collections and Preservation, University of Rochester River Campus Libraries)

"a place abounding with zoologists, taxidermists, osteologists and cast-makers, and specimens of a thousand kinds—humming with museum-making activities." Even more, Hornaday was captivated by the depiction of the taxidermy building. "The upper room in this building is a wonder to behold; hanging from the ceiling are hundreds of skins, including apes, monkeys, wolves, bears, hyaenas, lions, tigers, sloths, ant-eaters, armadil-los, buffaloes, deer, elk, moose, giraffe, yak, wild boar, peccaries." The list went on and on. The unmitigated praise for the skilled European prepara-tors and the descriptions of the buzz of activity in the workshops, the steady flow of specimens from the field and to major museums, and, most of all, that "marvellous 'skin-room,'" Hornaday recalled, "fired my blood."[3]

Bessey knew that Hornaday was desperate to learn taxidermy from "the best living teachers," so he advised his pupil to leave Iowa for Rochester.

That very night the young man wrote to Ward, requesting the opportunity to study at his establishment. "I have considerable knowledge of mounting birds and stuffed many specimens for the college museum last year," Hornaday wrote. "But my knowledge of the art is limited and it is my wish and determination to make a first-class taxidermist." Could Ward, he wanted to know, make a place for him? Ward's reply came quickly—but equivocated. "It did not say 'yes,'" Hornaday later remembered, "but it did not say 'no.'" Hornaday solicited letters of recommendation from Bessey and Iowa State's President Adonijah Welch. Ward was impressed by the references and, to Hornaday's delight, wrote to offer him a position as an assistant preparator. He instructed him to come to Rochester in November.[4]

Now, passing through the whale-jaw archway and into the courtyard, Hornaday was greeted by an ominous placard:

THIS IS NOT A MUSEUM!
But a Working Establishment,
Where all Are Very Busy.

Hornaday was unshaken; he told the first person he saw that he had come to see Professor Ward.[5]

WARD'S NATURAL SCIENCE ESTABLISHMENT

Henry A. Ward was born in 1834 on his family's farm, known as The Grove, outside Rochester. His grandfather Dr. Levi Alfred Ward Jr. had established a thriving financial enterprise there that began as a mercantile business and mail service and grew to include forays into insurance, banking, and real estate. Although Levi Ward had long ago given up on the medical profession, he believed that it was important to remain informed on unfolding discoveries in the natural sciences, and therefore encouraged young Henry after he began rock hunting. When Henry turned seventeen, Dr. Ward sent his grandson to Williams College to study geology. But Henry quickly realized he preferred the field to the lecture hall. He attended few classes and instead led other students on collecting trips into the adjacent countryside and as far away as the Connecticut Valley, Vermont, and New Hampshire. Once he even traveled alone to New York City to see the wonders on display at P. T. Barnum's American Museum.[6]

Finally overcome by wanderlust, Ward left Williams in 1852 and sailed to Paris with his friend Charles Wadsworth, whose father had offered to fund Ward's education at the École des Mines. While there, Ward also attended

Fig. 1.2. Henry A. Ward, circa 1865. (Courtesy of the Department of Rare Books, Special Collections and Preservation, University of Rochester River Campus Libraries)

morning lectures at the Jardin des Plantes and received private mentoring from Alcide Charles Victor Marie Dessalines d'Orbigny, professor of pale-ontology at the Muséum National d'Histoire Naturelle. At that very time, d'Orbigny was preparing his groundbreaking *Cours élémentaire*, in which he argued that paleontologists required training in zoology and botany, not merely in geology. D'Orbigny had traveled all over South America, collect-ing more than ten thousand specimens for the Muséum National between 1826 and 1833; he encouraged Ward to widen his interests to include all di-visions of the natural sciences.

Ward decided that if he was going to succeed as a scientific generalist and establish a specimen supply house for universities and museums, he had to embark on an expedition as ambitious as d'Orbigny's. He traveled first to Moscow, where he saw a preserved Siberian mammoth; back across Europe to Norway and Sweden; to London, where he met with geologist Sir Roderick Murchison; and to the west coast of Africa—all the while collecting specimens. By fall 1860, Ward had sent 170 large boxes back to Rochester for display at the university's Washington Hall. His grandson Roswell later described the astonishing array of forty thousand unpacked specimens as "ranging from tiny semi-precious stones to great blocks of basalt; from a series of glass models of the Kohinoor and other famous diamonds to a large selection of plaster casts of extinct mollusca, reptilia and mammalia."[7]

As Ward had hoped, his massive collection gave him instant credibility in the scientific community back home. He was made a member of the American Academy of Sciences and offered a professorship in natural history at the University of Rochester. Ward happily accepted the faculty appointment and quickly set to work designing three large teaching cabinets to represent the main branches of natural history—geology, zoology, and botany. But he felt he needed to undertake another trip to collect mounted specimens of exotic species for the new university museum he envisioned. Although the university's president, Martin B. Anderson, did not share his lofty goals, Ward, ever industrious, subsidized a return trip to Europe through public subscription.

From the specimen dealers of Paris, Ward "purchased about $6000 of vertebrate animals," including "the whale, porpoise, hippopotamus, elephant, tapir, wild boar, zebra, giraffe, ass, camel, llama, antelope, reindeer, great anteater, sloth, armadillo . . . seal, polar bear, lion, hyena, wolf, fox, vampire (bat), lemur, howling monkey, baboon, apes, chimpanzee, and GORILLA!"[8] He also purchased plaster casts of fossil skeletons of the extinct megatherium and glyptodon, which he envisioned forming the centerpiece of his museum. Impressed by the wonders Ward had collected, President Anderson relented and agreed to house the new cabinets in ten separate and cramped rooms on the upper floor of the university's main building, while two new buildings, which Ward dubbed Chronos and Cosmos Halls, were erected.

Ward saw everything coming together. "If the proper men come here and they work properly," he wrote to a friend, "Rochester University can be the center of educational science in America."[9]

WARD'S AS A TRAINING GROUND FOR TAXIDERMISTS

Less than a decade later, Ward's dream nearly came to an end. In late 1869, his son Charlie began to show an interest in taxidermy. To encourage the boy, Ward took him squirrel hunting and then, as evening drew near, walked him to Cosmos Hall to mount one of the specimens. Ward lit a candle and went into a storeroom to gather some tow to stuff the skin. In his hurry, he tipped over the candlestick, and the tow erupted in flames. Ward ran to Charlie and rushed him outside. By then, the wooden building was engulfed, and soon Chronos Hall was also on fire. The fire department responded, pumping the university cisterns dry, but it was too late. Chronos and Cosmos Halls both burned to the ground, taking with them most of Ward's collection. The damage totaled over $53,000—and Ward soon learned that the university had allowed its insurance on the collection to lapse. Worse still, the president refused to pay to rebuild.

Ward resolved to start his own business, independent of the university. He borrowed money from his grandfather and purchased land north of the campus.[10] By the summer of 1870, Ward had constructed the first building of the new Ward's Natural Science Establishment. Convinced that he could buck the country's unstable economy, still reeling from the Black Friday financial panic of September 24, 1869, he followed the first structure with additional buildings in rapid succession. Orders came at a steady pace, but the establishment was always just a step ahead of bankruptcy. Ward believed that he could ensure its survival by producing more and better scientific supplies than any of his rivals. To do this, he decided to hire a European workforce, as he felt there were no American institutions training museum professionals on par with those from France's Muséum National d'Histoire Naturelle or the Maison Verreaux. So Ward traveled again to Paris, where he hired the establishment's first preparators, Louis Charles Roch, an osteologist of small mammals, and Isidore Prevotel, a taxidermist.

Soon after, he began hiring Americans to assist and train with the two men. Frederic A. Lucas, Ward's first American preparator, was nineteen years old when he arrived at the establishment in January 1871. As the son of a sea captain from Plymouth, Massachusetts, young Lucas had been fortunate enough to travel on two separate voyages to the Far East and South America. His travels had fostered an interest in natural history, and by his second voyage he had assumed the position of ship's naturalist, collecting and sketching specimens, recording data, and practicing taxidermy—taught to him by an uncle who was a bird collector and amateur taxidermist. To his father's chagrin, Lucas sought to make a career in taxidermy. Augustus Henry Lucas, writing to Ward to introduce his son, explained that Frederic's life's ambition seemed "to be to skin snakes" and asked if Ward had "any use for such a specimen?" Ward replied, "Send him on!"[11]

With Lucas's knowledge of and interest in anatomy, Ward assigned him to the osteology workshop and set him immediately to mounting the skeleton of a pig for Louis Agassiz's collection of domestic animals at the Museum of Comparative Zoology at Harvard. But Lucas's presence in the osteology

Fig. 1.3. Frederic A. Lucas trained at Ward's Natural Science Establishment from 1871 to 1881. (Courtesy of the Department of Rare Books, Special Collections and Preservation, University of Rochester River Campus Libraries)

workshop angered Roch, who jealously guarded his methods of preparation and did not want apprentices looking over his shoulder. After Roch threatened to quit, Ward appeased him by giving Lucas more lowly tasks: mounting crustaceans, making crates, and packing for shipment "all sorts of objects—from elephants to humming-birds, plaster casts and skeletons."[12]

Within the year, however, the volume of orders coming in had increased to such a degree that Ward could no longer afford to ignore Lucas's skill with skeletal mounts. As Charles H. Townsend, a friend of Lucas's at Ward's, later recalled, "Mounting skeletons especially was of importance to [Lucas's] future. Sorting over the contents of a maceration barrel comprising two or three skeletons was the best possible training in comparative anatomy."[13] With Lucas back in the room, Roch made good on his threat and returned to France in disgust. Ward wasted no time in hiring his replacement, Jules

François Desirée Bailly, who had trained at the Maison Verreaux and had a broader range of experience than Roch. He was also the "first preparateur to come to America who could mount human skeletons."[14]

Bailly was the utter opposite of Roch. Happy and eager to train others in the preparatory techniques of osteology, he also had a quirky sense of humor—which he indulged by mounting frogs and other small animals in humorous human poses. As Lucas later reflected, "Bailly was a type of preparator rare in those days, in that he was quite ready to impart his knowledge and skill to anyone willing to devote time to its acquisition." Bailly became the establishment's first true instructor. "Under his supervision were trained a number of deft-handed preparators," Lucas wrote many years later, attributing the "excellent quality of skeletal preparations in [America's] great museums" to Bailly's willingness to share and develop ideas with his students.[15]

At about the same time, Ward hired Johannes Martens, a Dutch osteological preparator, who arrived from Hamburg, Germany. Ward's had become what Townsend called "a little community by itself, a polyglot community, including American, French, German, Swiss and Italian employees, each of whom was an expert in some branch of preparatory work such as taxidermy, osteology, or plaster work." This remarkable gathering of talent sparked the imaginations of many young naturalists who, like Townsend, were "trying to find themselves." Such had been the case with Lucas in 1871, and such was the case with William Hornaday two years later.[16]

In April 1873, when he received Hornaday's letter from Iowa, Ward was wondering how he would pay the workers to whom he already owed back wages, let alone hire another talented preparator. His greatest financial problem was getting institutions to settle their debts for cabinets delivered on credit during the previous year—and the problem grew worse with the Panic of 1873, when the U.S. economy sank into a deep depression that would persist for five years. Ward unknowingly had chosen to build his new enterprise on the shifting sands of devastating economic instability, and yet he managed to expand his business by taking on ever more ambitious jobs and demanding long hours of his preparators, even when their paychecks were uncertain.

Ward hired Hornaday at the modest wage of six dollars per week and put him to performing the establishment's most menial tasks: "piling empty boxes, scraping casts of fossils, digging drains, and [working] as Head Pumper in the water-logged cellar of Professor Ward's new and handsome house."[17] From this lowly perch, Hornaday watched Ward with wide-eyed admiration. He later described Ward in those years as having "the nervous energy

of an electric motor, the imagination and vision of Napoleon, the collecting tentacles of an octopus, and the poise of a Chesterfield." After three
months of basic labor, Hornaday reminded Ward that he had come to learn
taxidermy and was told to report the next morning to the "first floor front"
of the taxidermy building. As Hornaday remembered, "It was a thrilling moment when Frederic A. Lucas, the young scientific foreman of the works, led
me to the 'skin-room'—smelling of camphor and creosote—and showed me
a vast array of mountable mammal skins from all over the world." Lucas
told Hornaday to choose a skin to mount, but Hornaday could not decide,
so Lucas chose for him "a humble seal."[18]

Despite his youth, Hornaday had arrived at Ward's with an unusually
sophisticated background in natural history and taxidermy. The first specimen he had mounted for the Iowa State Museum was an American white
pelican collected on campus. When Professor Bessey presented Hornaday
with the skin, he also placed before him the five-volume set of John James
Audubon's *Birds of America*, instructing him to mount the bird according
to Audubon's artistic but lifelike representation—with the great bird posed
upright, its chest thrust proudly forward. Given this early training, Hornaday balked at the comparatively primitive methods of Isidore Prevotel and
Johannes Martens. "When I saw M. Prevotel make perfectly round stick-
like legs for monkeys, I knew that no wild animal had legs like those," Hornaday later remembered. "When I saw Johannes Martens stuff the bodies of
antelope and deer with oat straw rammed tightly, I secretly rebelled at those
also because I knew it was not just right."[19]

Yet, for the moment, Hornaday was grateful that the establishment
granted him access to such rare specimens. He marveled at how "the jungles of the tropics, the game-hunted mountains and plains, and the mysteri-
ous depths of the seas seemed to contest for the privilege of pouring in day
by day their richest zoological treasures." He viewed Ward's "like a signal-
station from which lines ran out all over the world," at which there arrived
daily boxes of "East Indian skins from Gerrard in London . . . a giant Lyre
turtle in the flesh, from New York . . . rough mammal skeletons from Paris
and a shipment of black iguanas and pink flamingo skins from Nassau."
During his first winter at Ward's, a shipment of American bison hides arrived from Wyoming, and, Hornaday later recalled, "the lot of shaggy hides
kept us on the jump for a week."[20]

A year at Ward's watching the constant influx of exotic specimens excited a desire in Hornaday to see live specimens in their natural habitats. As
he remembered it, "I heard the call of the wild," and after much convincing,
Ward agreed to train him as a field collector.[21]

HORNADAY IN THE FIELD

In October 1874, Hornaday sailed for Havana. Arriving in the midst of an insurrection, he was quickly ushered to the nearby Isle of Pines. On the island, he collected a Cuban crocodile, tree rats, large birds, and the skull of a manatee—killed one week earlier by a local fisherman—which Harvard's Museum of Comparative Zoology later purchased. From there, he traveled to Key West, where he collected tropical fishes, corals, shells, and sponges as well as Atlantic green and loggerhead sea turtles. In Arch Creek, near the head of Biscayne Bay, Hornaday collected—together with Chester E. Jackson, a young Wisconsin farmer and tourist in search of adventure—an American crocodile (*Crocodylus acutus*) measuring more than fourteen feet long. The discovery of crocodiles in North America caused a sensation and made a name for Hornaday in the scientific community. Ward's mounted the preserved skin, and the National Museum of Natural History purchased it for $250, later displaying it at the 1876 Centennial Exposition in Philadelphia.[22]

Ward must have been proud of his protégé, who, like him, seemed to relish not only the excitement and the danger of the hunt, but the thrill of participating in and contributing to science. Ward quickly organized a Caribbean expedition in the fall of 1875 for Hornaday and his newfound companion, Chester Jackson. Near Trinidad, on Huevos Island, Hornaday collected skins, eggs, and representative nests of the *guácharo*, or oilbird—a strange nocturnal species that, like bats, lives in caves and navigates by echolocation.[23]

Upon Hornaday's return, Ward sent him to mount a small exhibit for the Chicago Exposition of 1875. It was a minor fair compared with others of the day, but on his way home, Hornaday had a transformative chance encounter. He stopped to visit Ben Auten, an officer of the Battle Creek Sanitarium in Michigan, where Hornaday's mother had spent the last days of her life before her death nearly a decade before. Auten invited Hornaday to join him at a dinner party, where the young adventurer met a twenty-one-year-old high school teacher named Josephine Chamberlain. Hornaday was smitten. "She was a clear blonde, of a model fit for a figure of Diana, and her enunciation of pure English was a positive delight."[24] But Josephine was shocked by the brashness of this young man. He bragged about his discovery of the American crocodile and boasted that Ward had chosen him to leave soon on a dangerous expedition up the Orinoco River. Before parting from Josephine's company in Battle Creek, Hornaday promised to write her, but cautioned that local wisdom held that "when five men go up the Orinoco, only two return."[25]

In the early months of 1876, as Hornaday and Jackson ventured to mainland Venezuela and headed to the Orinoco in search of large mammals, the reality was less treacherous than promised, but tremendously fruitful. Together the two men collected a crocodile, a puma, a jaguar, capybaras, howler monkeys, an electric eel, and many species of birds. After Hornaday returned to New York, he began mounting these specimens to fill orders from Harvard's museum and for Ward's display at the Philadelphia Centennial Exposition.

In August, Ward proposed to send Hornaday on yet another collecting trip—a two-year expedition to Europe, Africa, India, Ceylon, and Borneo. Ward had recently met with Lewis Brooks, a local philanthropist who, in 1861, had given a gift of $5,000 to help Ward in establishing his natural history cabinet at the University of Rochester. Impressed by Ward's success, he wanted to fund instructional collections at two southern universities, alma maters of his Virginia relatives. Brooks wished to establish a large museum of natural history at the University of Virginia and a smaller natural science cabinet at Washington and Lee University. As he chose to remain anonymous, Brooks asked Ward to act as his liaison. He also asked for Ward himself to oversee the design of the University of Virginia museum building, now Brooks Hall. The generous budget and relative freedom would allow Ward to create the best collection yet, with even more exotic and rare specimens—and Hornaday was just the collector to acquire them.

Before he returned to the field, Hornaday asked Josephine to marry him and suggested that they meet in Philadelphia so they could tour the Centennial Exposition together.[26] This exposition marked a turning point for nineteenth-century taxidermy. Certainly, there were more obvious triumphs of American ingenuity displayed in Machinery Hall—Alexander Graham Bell's telephone, Christopher B. Sholes's Remington typewriter—but the quality of the taxidermy displayed was proof of a new American school. A new and effective method by which scientists could communicate their knowledge of the natural world to the public had emerged. Remarkably, it was not Ward's that pioneered this change, but lesser-known naturalists and taxidermists, such as Coloradoan Martha Maxwell.

Maxwell was exhibiting her taxidermy mounts in the Kansas and Colorado State Building—a far less prominent location than the Smithsonian's exhibit in the Government Building or the Ward's display in the Education Department of the Main Building. But Maxwell's booth would have captured Hornaday's attention, as it was superior to anything he had seen before. Elliott Coues, an influential American ornithologist, said of Maxwell's

Fig. 1.4. Martha Maxwell's booth at the 1876 Philadelphia Centennial Exposition.
(Author's collection)

exhibit: "It represented a means of popularizing Natural History, and making the subject attractive to the public; this desirable object being attained by the artistic manner in which the specimens were mounted and grouped together." Coues also believed that Maxwell's "skillful and faithful representation of nature" would eventually "come to be recognized as a means of public instruction." Given that Maxwell remained with her exhibit throughout the exposition, and that most visitors reported on their favorable interactions with her, it is possible that Hornaday not only saw Maxwell's taxidermy, but also spoke with her. Certainly, he would have been interested in her mounting methods, her artistic designs, and the naturalistic settings.[27]

Other revolutionary taxidermy exhibits that Hornaday would have seen included the U.S. National Museum's group of fur seals, Sweden's "The Dying Elk" and a reindeer pulling a sleigh, and Jules Verreaux's "Arab Courier Attacked by Lions." Verreaux's masterpiece, in particular, excited attention. Displayed by New York's American Museum of Natural History—which had purchased it in 1869—it was the first exhibit of its kind to be seen in America. While accurately depicting a hand-to-hand struggle between human and lions in northern Africa, it was generally viewed as too shocking and melodramatic to be associated with the serious work of naturalists—yet the taxidermy was so accomplished that the AMNH had not only chosen to acquire the exhibit, but to showcase it as a highlight of its fledgling collection. Likewise, the Swedish government exhibits depicted remarkably realistic taxidermy mounts, but the animals were presented only in relation to human interactions—as big game hunted for food or as domesticated work animals.

By contrast, the group of fur seals mounted by Julius Stoerzer for the National Museum's display was a quiet but impressively executed arrangement of sixteen fur seals lazing on a base of artificial rocks. Certainly, Hornaday would have seen this exhibit when he took Josephine to see his American crocodile and other mounts he had prepared for the Smithsonian display. Stoerzer's group was notable for what *Forest and Stream* considered its "lifelike attitudes"—depicting a single bull with his harem of cows and pups, "rolling about in play or 'drawing in milk' from their mother's udders."

Fig. 1.5. Jules Verreaux's "Arab Courier Attacked by Lions." (Carnegie Museum of Natural History)

The magazine proclaimed the exhibit a "masterpiece of taxidermy" and called Stoerzer "the most thoroughly trained and scientific workman in the country." Unfortunately, the gifted taxidermist had died, at the age of only thirty-four, shortly before the exhibit opened.[28]

Although it is impossible to know exactly how Hornaday reacted to these exhibits, it is clear that Maxwell and Verreaux had an influence on the ideas Hornaday developed over the next two years while collecting specimens for Ward's. After leaving the exposition, Hornaday traveled directly to New York City, where he and Ward set sail aboard the steamship *Bolivia* for the port of Londonderry. Hornaday visited every museum from Ireland to Paris, but none impressed him more than the British Museum, which in his view was "the most complete of any of its kind in existence, and always will be. It outranks all other museums. . . . There is not now, and there never will be, even in boastful America, another museum which can even be compared with it as to size and scientific completeness."[29]

For the next two months, Hornaday and Ward made their way across Europe to Egypt. At the Red Sea, Ward turned north and headed back to Germany, while Hornaday continued eastward to India. "Well, Hornaday," Ward said, bidding his protégé farewell, "there's no knowing whether or not we shall ever see each other again." He meant the words sincerely: the places Hornaday was venturing to were remote and dangerous then, particularly his eventual destination in Borneo. Still, Ward's expectations were high: he instructed Hornaday to collect "skins and skeletons of elephants, Indian bison and elk, orang-utans, gibbons, monkeys of all species, two or three tigers if practicable, and every species of crocodile procurable." Hornaday was an experienced traveler now. He packed a perfect outfit, complete with compass, guns and ammunition, measuring tape, skinning knives, and other tools used in the preparation of skins in the field: arsenic, labels, and specimen boxes, and one Agassiz copper tank in a wooden box, used for preserving specimens in alcohol.[30]

Upon arriving on the subcontinent, however, Hornaday's adventure was slowed by bureaucratic delays and the difficulties of working in the dense Indian forests. It took him weeks to secure permission to hunt male elephants in the Annamalai Forest and another five months to track them down. To make matters worse, when Hornaday finally did fell a good-sized male, he had extreme difficulty clearing an area in the thick jungle in which to work. The carcass spoiled before he could prepare the skin, and Hornaday was only able to salvage the skeleton. Finally, in November 1877, he reported to Ward that he had shot and killed another large individual—but there was a problem. "I'm sorry to say it's a female, and stood 9 ft 10 in. at

Fig. 1.6. William T. Hornaday's rendering of skinning the Indian elephant in his book *Two Years in the Jungle*. Note that tusks have been added to the illustration to disguise the sex of the specimen. (Author's collection)

the shoulders, as I remember. It fell in about the worst jungle. . . . it was that or nothing, and I had grown desperate at last. . . . Please don't let it be known in the papers that this is a female, for then I would be in hot water if it ever got back here. My permission is for a *tusker* you see." In his memoir of the trip, *Two Years in the Jungle*, Hornaday wrote of his successful hunt of a "noble tusker," even including a fabricated length for the specimen's nonexistent tusks. To complete the deception, Hornaday published a drawing depicting the skinning of a male elephant.[31]

In a letter to Alexander Agassiz, at Harvard's Museum of Comparative Zoology, Ward offered "a full-grown Indian elephant," making no mention of its sex, and Agassiz eagerly agreed to purchase it. While preparing the skin and skeleton of the elephant in the field, Hornaday found "a *foetus*, quite well developed," and he "saved the skin and skull of it." Writing to Ward, Hornaday expressed a macabre desire to "put in the first bid" for the animal: "The skull doesn't amount to a great deal, but I have a hankering after that little skin to stuff for a library ornament. I must have *some* trophy of my elephant shooting."[32]

Hornaday and Bailly later mounted "The Cambridge Elephant" in 1880. They probably used the tusks from the skull of the male Hornaday had shot earlier in the collecting trip to turn the female into "a full-blown tusker."[33] And Ward not only kept Hornaday's secret, but allowed him to keep the fetus.

THE ARRIVAL OF WEBSTER AND TOWNSEND

While Hornaday was hunting elephants in India, Ward had learned of an exceptional bird taxidermist from Troy, New York, named Frederic Smith Webster. Since 1867, Webster had been designing bird groups and photographing them using the wet plate process to produce stereoscopic pictures, which he printed for "school and family use." In 1876, George Burritt Sennett—a businessman and naturalist from Erie, Pennsylvania, who was preparing for a trip to the lower Rio Grande and Mexico to make a study of Texas birds—was shown some of Webster's stereoscopic views. Sennett wrote to Webster suggesting that he should mount the birds collected on the expedition, and "he induced me to accompany him," Webster later remembered, "with expenses paid but with no salary."[34]

In late January 1877, on his way to meet Webster at his studio in Troy, Sennett stopped to see the Ward's establishment in Rochester. Ward was then in Africa collecting, but his new partner, Edwin Howell, was there to greet Sennett. "Saw their very complete work shops employing 18 hands," Sennett recorded in his notes. "They have few birds and do not wish them. They have, I should judge, a hundred Rocky Mt. Sheep just received and skins by the hundred of other animals." Sennett went straight from Ward's to see Webster, whom he found to be "not only a good taxidermist but an observing and careful naturalist and scientist." Webster and Sennett spent the next day together at the New York State Museum in Albany, visiting with its director, James Hall, and the entomologist J. A. Lintner. After leaving Webster, Sennett completed his New York trip by stopping at the AMNH, where he met museum superintendent Albert S. Bickmore and his assistant, Joseph B. Holder, who "gave me much attention and made me interested beyond expression in their work and collections."[35]

For two months, as they traveled across Texas and northern Mexico, Webster heard Sennett's stories of the wonders of the AMNH and the remarkable natural history establishment that supplied most of its specimens. When the two men returned to Erie in May 1877, Webster mounted a number of the birds collected on the expedition and catalogued the entire collection of skins. Although Sennett believed that Ward's didn't seem in-

Fig. 1.7. Frederic S. Webster. (Library of Congress)

terested in bird taxidermy, he began to wonder if there might not be a place
for such a skilled taxidermist, regardless of his area of specialty. With Sennett's encouragement, Webster sent Ward a letter and twenty-four stereoscopic photographs of what Webster called his bird "group-sets." Webster
later remembered, "To my great surprise and even greater satisfaction, three
days later, I received a complimentary letter in reply from dear old Professor Henry A. Ward, closing with the words, 'You may come on at once!'"[36]

Arriving in December, Webster waded through deep snow deposited
by a recent storm before finally passing through the whale-jaw archway of
Ward's establishment. He entered the large main building, where he found
"the Wizard of Ward's hard at work in his crowded sanctum." Ward guided
Webster through the snowdrifts piled around each of the seven workshops
and laboratories. Together the men entered a two-story building—the taxidermy workshop. The room was cold and smelled of damp straw, "in spite of
its 'big-bellied' and red-hot coal-burning stove." The frost-tinted windows

Fig. 1.8. The taxidermy workshop at Ward's Natural Science Establishment, as drawn by Frederic A. Lucas for *Ward's Natural Science Bulletin*, 1883. (Courtesy of the Department of Rare Books, Special Collections and Preservation, University of Rochester River Campus Libraries)

shined a diffuse light onto a partially mounted horse—General Philip Sheridan's war-horse, Winchester—being worked on by Martens.[37]

They continued up a staircase to another room with better lighting, where Bailly was mounting several birds. In the back of the room lay a pile of about two hundred skins of "trogons, birds of paradise, impeyan, tragopan, and peacock pheasants, scarlet and bronze ibises, and innumerable smaller birds from many foreign lands." Ward explained that the specimens had been discarded by Bailly because they were too difficult to mount. Webster countered that if the specimens were unmountable, it was not because of the condition or quality of the skins, but because "the man did not have the proper method to apply." When he boasted that he could save the skins and mount them himself, Ward responded, "Well then, that's your particular job and begin as soon as possible." Agassiz later purchased the specimens Webster mounted from that pile for the Museum of Comparative Zoology.[38] During his tenure at Ward's, Webster mounted numerous specimens, most for Harvard's museum, but also for Amherst College—including about a hundred American birds collected by John James Audubon, which had been sold to Ward by his daughter Mary Audubon. Another preparator at Ward's several years later recalled that when Webster arrived in Rochester, he

> brought with him such degree of both artistic and mechanical skill in mounting birds, and such a perfect knowledge of their forms and habits, that no European taxidermist has ever been able to add one iota to his professional ability. Mr. Webster is wholly an American taxidermist, with a true genius for bird work.[39]

Although Ward did not consider the establishment an educational institution, word continued to spread that it was a training ground for this wholly American school of taxidermy—where the best preparators in the country gathered, practicing a new scientific discipline that was producing mounted specimens superior to any seen in natural history museums throughout Europe.

In 1879, Ward hired another promising young taxidermist who had heard the news of his establishment. Charles Haskins Townsend was raised in the village of Beatty, Pennsylvania, twenty-five miles outside of Pittsburgh. The first book he ever read about birds, "a happy discovery" in his family's crowded library of theological literature, was Ezekiel Holmes's *Birds Injurious to Agriculture* with thirty-two full-page woodcuts of Audubon's birds, which Townsend used as a field guide. Like Hornaday, Townsend attended the Centennial Exposition in Philadelphia, where for the first time he saw thousands of taxidermy specimens, but it was the migrating flocks of passenger pigeons, which darkened the skies over his hometown, that inspired him to learn how to mount birds. In fact, his first attempt at taxidermy was a pair of passenger pigeons that he killed in the "scattered oaks" near his house with a single-barreled muzzle-loader borrowed from a neighbor. In an autobiographical sketch written fifty years later, Townsend recalled the summer of 1876 as the last that he observed large flights of the dwindling species. Three years later, he wrote to Ward and told him how his establishment's exhibit in Philadelphia had inspired his work, and the old man responded with characteristic enthusiasm, inviting Townsend to come on as an apprentice at the establishment, where he would learn the art of taxidermy from the world's best.[40]

THE GROUP IDEA AND THE DESIRE TO PROFESSIONALIZE

When Hornaday, sun-bronzed and lean, stepped off the train at the station in Rochester, New York, in 1879, he was greeted by his smiling mentor.

"Well, Hornaday," Ward began. Hornaday grinned back and finished the sentence the two men had parted on two years before: "There's no knowing whether or not we shall ever see each other again."[41] As they made their way back to the shop, Hornaday told Ward an amazing story about hunting orangutans in Borneo.

For days, the native Dyaks had been paddling him in their long dugouts to spots on the Sadong River overhung by durian trees; when Hornaday spotted dark shapes moving in the upper boughs, he would level his shotgun and fire. The collecting had grown ruthlessly efficient, even routine—on one particular day felling seven orangutans—but then one day the Dyaks brought Hornaday a remarkable old male. "He bore the scars of many a hard-fought battle," he later remembered.

> A piece had been bitten out of his upper lip, and the lower lip also had been bitten through; both middle fingers were off at the second joint, leaving mere stumps; the third right toe had disappeared from the same cause; the fourth left toe and both the great toes had been bitten off at the end; one finger was quite stiff and misshapen from a bite, and, to crown all, he was actually hump-backed, caused, as I found on dissecting, by some violent injury, possibly a fall. He had evidently been a regular prize-fighter in his day, a first-class desperado.[42]

Hornaday explained to Ward that picturing the fierce conflicts of this "battle-scarred hero" made him begin to imagine mounting something more than an imitation of the usual pose for large primates—standing, one hand gripping an upright, leafless branch. Instead, he envisioned a group of orangutans, mounted together in dramatic poses, swinging, as they had in life, from one leafy bough to another, while two males at the center are locked in deadly conflict.

Hornaday handed Ward a sketch for a proposed group of five orangutans, which he later described in meticulous detail:

> A pair of immense and hideously ugly male orang utans fighting furiously while they hung suspended in the tree-tops. The father of an interesting family was evidently being assailed by a rival for the affection of the female orang utan, who, with a small infant clinging to her breast, had hastily quitted her nest of green branches, and was seeking taller timber. The nest which she had just quitted was an accurate representation of the nest constructed by this great ape.

In the middle of the group, and at the highest point, was another nest in the top of a sapling, on the edge of which another interesting young orang utan—a production evidently of the previous year, was gazing down with wide-eyed wonder at the fracas going on below. The accessories to this were so designed and arranged as to represent an actual section of the top of a Bornean forest, at about a height of about thirty feet from the ground, representing the natural trees, with leaves, orchids, pepper-vines, moss, and vegetation galore.[43]

Ward listened intently to Hornaday's proposal—but, at first, he was reluctant. He had been without Hornaday's help in the shop for years now, and the trip had come at considerable cost. Not only would a group like the one Hornaday proposed take a substantial amount of time to prepare, but such a large group had never before been sold to an American museum. Museums already complained that the establishment's prices for simple, single taxidermy specimens were too high, and the number of animals in complex poses in Hornaday's proposed group would dictate an unprecedented price.[44]

Hornaday argued that this obstacle would be overcome by the inclusion of the dramatic scene of conflict between the two central adult males. He admitted that the design "was highly suggestive of the methods adopted by my European rivals to secure attention to their work, or, in other words, it was a trifle sensational,"[45] but he thought it would entice some museum to make the purchase—just as the dramatic action of "Arab Courier" had made it irresistible to the American Museum. "After considerable hesitation," Hornaday later recalled, "Professor Ward finally decided to let the experiment be tried."[46] His one condition was that Hornaday agree to spend only two months completing the work. He wanted the group complete in time for the meeting of the American Association for the Advancement of Science (AAAS), to be held in Saratoga Springs, New York, in August.

After many long days and sleepless nights, Hornaday met his deadline; he shipped the pair of fighting males more than two hundred miles by rail to Saratoga Springs. At the AAAS meeting, Hornaday presented his field observations on orangutans in a lecture titled "On the Species of Bornean Orangs, With Notes on Their Habits." As he began his talk, he unveiled the group, which he dubbed "A Fight in the Tree-Tops," to illustrate the nature and habits of orangutans. The assembled scientists were overawed—but less by the dramatic composition, as Hornaday had imagined, than by the quality of Hornaday's artistry and precise attention to anatomy.

Fig. 1.9. The two clashing male orangutans from "A Fight in the Tree-Tops," mounted by William T. Hornaday for the AAAS meeting in Saratoga Springs, New York. As pictured in the *First Annual Report of the Society of American Taxidermists*, 1881.

George Brown Goode, assistant secretary of the Smithsonian, was among those in the audience. He later reflected that Hornaday's group was "extremely spirited and had all the qualities of good workmanship and permanence." Goode considered Hornaday's orangutans superior to the "figure groups" of Verreaux, which he thought embodied "false ideals." In fact, he was so impressed that he offered Hornaday a position at the U.S. National Museum as chief taxidermist—a position vacant since the death of Julius Stoerzer. Hornaday declined for the moment, as he felt he owed Ward at least two years in return for his time abroad.[47]

Albert S. Bickmore, superintendent of the AMNH, who also attended the meeting, made Hornaday a similar offer. Hornaday again declined. Undeterred, Bickmore convinced trustee Robert Colgate to commission a similar group of orangutans from Ward's on behalf of the new museum on Central Park. Hornaday later contended that the price of $2,000 that Ward had placed on "A Fight in the Tree-Tops" had "prevented its immediate sale," but Colgate had offered a sum of $1,500 for the new group.[48] It was more likely that the group's dramatic composition—which Hornaday had imagined would be its selling point—had instead discouraged Bickmore and other potential buyers. Bickmore obviously admired Hornaday's skill, but the AMNH already had a lurid crowd-pleaser in "Arab Courier."

By Hornaday's own description, the group he mounted for the American Museum, called "The Orang Utan at Home," was "similar in composition but of a very different design." As he explained, "This group represented the orang at home—a perfectly peaceful scene in the top of the Bornean forest. It included five orang utans, of various sizes and ages, feeding on durions, sleeping in a nest, climbing, sitting, and swinging."[49] Ironically, the old battle-scarred male that had inspired Hornaday to imagine the fighting scene was mounted in quiet repose for Bickmore's group, and Hornaday later mused, "Alas! for him, his fighting days are over, and he now peacefully sits on the branch of a tree in the American Museum of Natural History, quietly eating a wax durian."[50]

In response to potential buyers' reluctance to pay so high a price for "A Fight in the Tree-Tops," Ward reduced the group to the two central figures and lowered the price to $800. The entire setting would not be restored until August 1883, when Hornaday—by then in the employ of the Smithsonian—convinced the U.S. National Museum to purchase the whole group from Ward. He then "partly reconstructed" the piece for exhibition in the Hall of Mammals. During that same month, the *Washington Post* reported that the group, "mounted in the highest style of the taxidermist's art," was soon to be installed in a large glass case in the southwest corner of the museum

Fig. 1.10. William T. Hornaday's "The Orang Utan at Home," mounted for the
American Museum of Natural History. (Image #37605, American Museum of
Natural History Library)

building.[51] For nearly a century to follow, Hornaday's two orangutan groups
would be the way that most Americans on the Eastern Seaboard encountered these Asian great apes in the flesh.

More immediately, Hornaday's success gave hope to the preparators at
Ward's, many of whom had grown dissatisfied with the state of taxidermy
in America and Europe; they regarded Hornaday's renown as an opportunity to overthrow the secrecy surrounding old-fashioned methods of preparation, which they believed was preventing taxidermy from reaching the
highest levels of achievement. Hornaday, for his part, saw that this group of
taxidermists who had been assembled by Ward in his absence—particularly
Frederic S. Webster—had the requisite skills to formally establish a new
American school. Webster later remembered: "We were of about the same
age, had similar inclinations, and formed a firm friendship from the very

start. He saw what I was doing and had done, and we got our crusading minds together."[52]

On March 12, 1880, seven employees of Ward's—four Americans and three Europeans—gathered to discuss the formation of the Society of American Taxidermists. Soon after, the four Americans were chosen to hold office: Webster as president; Thomas W. Fraine as vice president; Hornaday as secretary; and Lucas as treasurer. Though Jules Bailly and Johannes Martens had laid the foundations of taxidermy at Ward's, it was important to the group that the SAT be a society to professionalize American taxidermists and thus establish a new, purely American style. By committee, the group drafted and adopted a constitution. The mission of the society would be "to promote intercourse between those who are interested in the art of Taxidermy in various parts of America, to encourage and promote development of that art, and to elevate it to a permanent and acknowledged position among the fine arts."[53]

Fig. 1.11. The Taxidermy Department at Ward's Natural Science Establishment, circa 1880, as pictured in Frederic A. Lucas's *The Story of Museum Groups*. Society of American Taxidermists members are marked with asterisks. *Standing, left to right*: Frederic S. Webster,* Harry L. Preston, Edmond Gueret, Arthur B. Baker,* Robert Koehler, Frederic A. Lucas,* J. William Critchley,* Frederick W. Staebner,* E. Mirguet; *seated, left to right*: Nelson R. Wood,* Isidore Prevotel, Charles E. De Kempeneer,* William T. Hornaday,* Johannes Martens,* Jules F. D. Bailly.* (Author's collection)

To carry out this mission, the founders decided to hold a competitive exhibition in concert with the society's annual meetings. Taxidermists would vie for medals and certificates awarded by three judges, professionals in the field of zoology. To announce their organization, they mailed five hundred copies of the society's constitution and three hundred circulars "to every taxidermist in this country whose name and address could be obtained." Membership was open to both amateurs and professionals upon written recommendation of one member, nomination by the executive committee, and election by a majority of members.[54]

To induce others to join, the society elected many prominent scientists and patrons of taxidermy as honorary members. The first were Fred T. Jencks of Providence, Rhode Island, and W. E. D. Scott, curator of the Princeton College Museum. Both men became active members of the society. Later, other honorary members were added to the list, including Henry A. Ward; J. A. Allen of Harvard's Museum of Comparative Zoology; and Spencer F. Baird, George Brown Goode, and Dr. Elliott Coues, all of the U.S. National Museum. The nominations were accepted with words of "sympathy and encouragement," as all understood the difficulty of the undertaking.[55]

Ward, in particular, was pleased to hear that his employees endeavored to raise the position of taxidermy to a place among the fine arts, thus "conferring a solid boon on the science of zoology, besides winning for yourselves an enviable reputation as artists." Goode, having just been to the Fishery Exposition in Berlin, had been able to compare the work of Ward's taxidermists on display there with the work of the Germans[56] and found it to be "fully equal to the best examples to be found in the Museum at Berlin." Impressed by its members' abilities as taxidermists, he did not doubt that the society would "do much towards stimulating study and experiment, and that by its means mechanical and artistic perfection in work of this class will be more nearly approximated than ever in the past." Goode's words would soon prove prophetic.[57]

CHAPTER TWO

"Breathing New Life into Stuffed Animals":
The Society of American Taxidermists

The rise of American taxidermy to a level with the other fine arts thus
far is a chapter of unwritten history. It is probable that not more than a
score of persons now living know the real story of the Society of American Taxidermists, and the revolution that it wrought.
—William T. Hornaday[1]

On the night of December 14, 1880, 350 special guests of the Society of
American Taxidermists filed into the gaslit exhibition hall at 69 State
Street in Rochester, New York. Excitement had been building all week as
the society unveiled displays in local shop windows, including an elaborate
exhibit that had drawn crowds of the curious outside C. E. Furman's cloth-
ier. As visitors now reached the top of the exhibition hall stairs, they were
greeted by a pair of mounted great cats—to the right "a splendid lion," its
head turned and mouth open wide, and to the left one of the largest tigers
ever exhibited in America, nearly ten feet from nose to tail, that the society's secretary, William T. Hornaday, had shot in India.

Inside, the hall was thirty feet wide and more than a hundred feet long,
but the room was packed tight with "birds, beasts, reptiles, fish, insects and
about everything that can be preserved by the taxidermist's skill." One wall
was covered with examples of decorative taxidermy: deer and elk heads,
panels of dead game, feather fire screens, and a variety of owl species. Bird
groups in elaborate cabinets were hung chockablock on the opposite wall.
On an elevated platform in the center aisle, toward the back of the hall,
stood the single mammal specimens, described by the *New York Tribune* as
including "a noble American bison, a mountain sheep of great beauty standing proudly on a rock, an antelope, a lioness, a black bear, a fur seal and her
young, and various smaller animals."[2]

41

Most specimens were mounted in the stiff poses visitors were accustomed to seeing in museums, but there were a few mounts that departed dramatically from that convention. These exhibits, including Hornaday's pair of fighting orangutans and a group of three American flamingos mounted by the society's president, Frederic S. Webster, generated tremendous excitement. For most in attendance that night, it was not the first time they had seen such creatures, but it was certainly the first time they had viewed mounted specimens in lifelike poses and arranged in naturalistic settings.

FIRST ANNUAL EXHIBITION

Following the successful private showing of the previous night, the society's first annual exhibition opened its doors to the public on December 15, 1880, and remained on display for a week. The society had intended the exhibition to last only four days, but public enthusiasm swelled as word-of-mouth and positive press coverage, both in Rochester and in New York City, made certain its "character and merits" were widely known. "A disappointed visitor has not yet been seen, and the attendance steadily increases," the *New York Tribune* reported. "The success of the society is now assured and its permanence as a national organization established."[3]

It is easy to understand why the public was so enthralled. Exhibited on a small round table near the entrance was "An Interrupted Dinner." The group, mounted by Frederic A. Lucas, the society's treasurer, depicts a red-tailed hawk that has just killed a ruffed grouse, but before the meal can be devoured, "a goshawk swoops down upon him with outstretched talons to seize the quarry." The hawk is on its back, protecting its prey with its left wing, its bloody beak and talons raised defensively. Lucas cleverly mounted the goshawk on a brass standard hidden in the tail, giving it the appearance of hovering in midair. Hornaday described the exhibit as "the most striking table group I have ever seen."

The exhibition included many more artistic bird groups, including species "from the bald eagle to the humming bird . . . arranged with natural scenery and background effects to represent the haunts and habits of the birds." Hornaday had designed a mixed species group of his own, called "Does Your Mother Know You're Out?" (earlier titled "A Mutual Surprise"), set along a riverbank in the tropics, in which a scarlet ibis comes upon an alligator newly hatched from its egg.[4]

Perhaps the strangest displays were the grotesques, which the exhibition catalogue promised would "furnish an endless amount of amusement to old and young." These popular novelties of the time—which featured kittens

Fig. 2.1. View from the entrance of the First Annual Exhibition of the Society of American Taxidermists in Rochester, New York. Webster's flamingo group is seen in the foreground. Frederic A. Lucas's "An Interrupted Dinner" is on the table to the right of the flamingos. William T. Hornaday's orangutan group is just visible behind and to the left of the flamingo group. Published in the *First Annual Report of the Society of American Taxidermists*, 1881.

"making love"; frogs dueling, drinking, smoking, and fishing; and squirrels playing cards and dominoes—were placed on the floor here and there throughout the room. Most of the grotesques were the product of Jules Bailly's bizarre imagination and idiosyncratic sense of humor. A particular favorite among his groups, labeled "The Taxidermist," poked fun at Webster, who was depicted as a frog sitting at a table mounting a hummingbird.[5]

The undisputed focal points of the exhibition were positioned at the end of the hall: the pair of orangutans mounted by Hornaday and the group of flamingos mounted by Webster. Hornaday's exhibit was clearly the more dramatic and startling of the two, and Webster found it "all too real to be pleasing to children and sensitive persons," but conceded that "like all gruesome things it attracted attention."[6] It obviously succeeded in winning the attention of the judges: it was awarded the silver medal, the top award, for best piece in the entire exhibition.

In contrast to the clashing orangutans, Webster called his bird group "The Flamingo at Home"—its name mimicking Hornaday's second group, "The Orang Utan at Home." The exhibition catalogue described Webster's group in careful detail:

> In the shallow water, near the edge of a tropical lagoon, a female Flamingo has built her elevated nest of mud and grass, and in a half standing posture is covering her eggs. This nest is modeled according to the description and measurements given by Audubon. At the left of the nest, a stately male Flamingo on the bank is stepping into the water, while on the right another large male bird is stooping down, intently watching a small turtle which can just be discerned at the bottom of the water. The accessories, a dwarf palmetto and aquatic plants, are purposely few in number, and many desirable features in color have been omitted for the sake of preserving the entire naturalness of the surroundings.[7]

Months earlier, when Webster informed Ward that he intended to mount a group of flamingos for the society's first exhibition, Ward expressed doubt. Webster recalled the conversation years later: "'Where are you going to get the birds?' 'Why, Professor! You have several of them!', I reminded him. 'Ho-ho! that's it, is it? Well, you will have to interest me more than I am at the present moment.'"[8] Ward was confident in Webster's abilities as a taxidermist, as he had already mounted what Ward called a "pretentious" group of platypus, but he had reservations about his knowledge as a naturalist—whether his mounts would hold up to the scrutiny of trained ornithologists. After much convincing, Ward agreed to sell Webster three African flamingo skins at the price of seventy-five dollars. Webster admired the beauty of the African species above the American, for "their rich pink tints which contrast with the chalk-like white of the rest of the body and the faint rose flush of the long neck."[9] Webster planned an elaborate habitat that he hoped would "influence Professor Ward to advance the educational value of museum exhibits by developing habitat bird groups."[10]

To answer Ward's concerns about his scientific knowledge, Webster pored over the description of the species in John James Audubon's *The Birds of America*—but there was a problem. In outlining the flamingo's nesting habits, Audubon had relied on a secondhand account by the English explorer William Dampier, who wrote in 1699 that the female covered its nest by "standing in the water on one foot and supporting its body on the nest."[11] In order to incubate eggs in this way, the flamingo would have had to straddle its built-up nest and lower its body onto the rim of the mud walls. As

Fig. 2.2. Frederic S. Webster's flamingo group. Published in the *First Annual Report of the Society of American Taxidermists*, 1881.

Webster recalled, "Common sense told me that no flamingo could strain its anatomy by resting for hours on the sternum without the support of its legs." He reasoned that the flamingo instead would have to step inside its nest and lower its body by folding its legs underneath. But Ward demanded that Webster not "fly in the face of authority" and insisted that he mount the specimen in accordance with Audubon's description. In the end, Webster compromised, mounting the female straddling its nest, but with both feet touching the ground. He was later vindicated when ornithologist C. J. Maynard observed and reported in 1884 that flamingos do not "straddle" their nests. But it would be another twenty years before the ornithologist Frank M. Chapman—returned from the Bahamas with photographic evidence of American flamingos sitting on their nests with their legs folded under their bodies—mounted them in the correct position for the American Museum in a habitat diorama, called "The Flamingos of Andros."[12]

The society's founding members were proud of the flamingo group and fully believed that it would win a medal. At the general meeting, society members chose J. A. Allen; Joseph B. Holder, now director of the AMNH; and W. E. D. Scott, who had been elected the new president of the society, to judge the entries. Two days later, the judges submitted their sealed report, but their selections were not made known until near the close of the exhibition, and after they had taken leave of the city. To everyone's surprise, the judges resisted the consensus opinion of the taxidermists and chose not to acknowledge the flamingo group; it was not even awarded a certificate of honor.[13] Instead, they awarded the bronze medal for second-best piece in the exhibition to another specimen mounted by Webster, a conventionally presented wood duck. Webster was no doubt disappointed, but remembered that the duck was "a real classic."[14] Hornaday was less forgiving, referring to the specimen as "a silly solitary little wood-duck on a 25-cent pedestal of black walnut."[15] Society members were outraged and demanded that the judges provide a justification. According to both Webster and Hornaday, the judges explained that the flamingo group was "an attempt to attain the unattainable."[16] Accustomed to scientific specimens displayed in endless rows, they rejected the idea that museum taxidermists could accurately represent an animal's natural form and re-create its habitat in the display case. They must have recognized that the taxidermists intended the flamingo group to serve as a model for a series of similar habitat groups that Ward's would sell to natural history museums.[17]

The society, intent on pursuing the idea of the habitat group, responded by resolving to appoint two artists and only one scientist to judge the second exhibition the following year. While Hornaday and Webster worried about the future of the group idea, Lucas—the most practical and scientific minded of the group—argued in his treasurer's report that "lest the society should come to be regarded as a merely local one,"[18] the next year's exhibition should be held outside of Rochester.

The first exhibition closed at 10:00 p.m. on December 21. The *Rochester Democrat and Chronicle* reported that the society "demonstrated in a manner eminently satisfactory to all concerned that they will uphold every enterprise that has for its fundamental principle the advancement of science and general knowledge."[19] The reporter had probably spoken to Hornaday and Webster, who would have emphasized the scientific merit of their work—regardless of the judges' opinion that habitat groups, particularly for birds, were not scientifically legitimate. Despite the setback, the officers continued to emphasize an artistic approach to taxidermy, encouraging members to mount specimens in dynamic postures and in re-creations of their natural habitats. Society members hoped that by creating a desire in the general public to see these habitat groups, they eventually would convince natural history museums of the instructional value of habitat group displays.

SECOND ANNUAL EXHIBITION

Hornaday, Lucas, Webster, and William Critchley—another taxidermist from Ward's—arrived in Boston on December 10, 1881, and began unpacking the exhibits for the society's second annual exhibition at Horticultural Hall. Only two weeks prior, Reverend William Elgin, one of the society's original members, had traveled from New York to Boston to locate and secure a hall. Built shortly after the Civil War by the Massachusetts Horticultural Society, Horticultural Hall was the perfect venue. The exhibition was scheduled to open its doors to the public on December 12 and run through December 21. The society was working against a tight deadline. Although others soon arrived to help with the installation, the opening was delayed for two days. The exhibition was trumpeted to the public in daily advertisements in six Boston newspapers—each declaring, "Interesting for everybody!" Announcements were distributed around the city on a hundred large three-sheet posters and two thousand half-sheets that were hung in storefront windows.

The society's second general meeting was held in the lecture room of the Boston Society of Natural History on the evening of December 13. After President W. E. D. Scott gave the opening address, which was laudatory of the society's accomplishments, two technical papers were presented by Hornaday, "On the Uses of Clay as a Filling Material" and "Mounting Fish for the Cabinet." After Hornaday, Lucas presented "A Critique on Museum Specimens." Lucas's speech was an especially bold condemnation of natural history museums, where, he believed, visitors were at first "a little dazzled by the number of animals but as this feeling wears away we notice that there is somehow a certain air of monotony about them all."[20] He especially disliked the uniformity of poses: "Nine tenths, or more of the Carnivores have their mouths wide open, and are trying to look fierce without having adequate cause," and "birds we find arranged in serried ranks and look as if the greater part had been turned after a model by an eccentric lathe."[21] Though Lucas placed some blame on taxidermists, he argued that they were only producing what museums required. He then read from a letter sent by Elliott Coues, the society's first honorary member and honorary curator of mammals at the U.S. National Museum, in which Coues stated flatly that "museum birds are for study, and 'spread eagle' styles of mounting, artificial rocks and flowers, etc., are entirely out of place in a collection of any scientific pretensions or designed for popular instruction."[22] Yet clearly Coues was conflicted. Only a few years earlier, in a published review of Martha Maxwell's exhibit of Colorado mammals at the Centennial Exposition in Philadelphia, he had argued that her naturalistic taxidermy and methods of display in habitat groups "represented a means of popularizing Natural History" and would "come to be recognized as a means of public instruction."[23] Despite their value in educating the public and popularizing natural history, however, Coues steadfastly rejected the idea that mounted specimens in habitat groups could have scientific value.

It seems that the society's stated mission "to elevate it [taxidermy] to a permanent and acknowledged position among the fine arts" was not its only mission.[24] In fact, it was increasingly clear that the society had an important unstated goal: to have naturalistic or artistic taxidermy serve a dual purpose as both an object of fine art—recognized as such for the taxidermist's ability to render the animal in death as it was in life—and as a museum specimen valued as an educational object that could be used to inform the public about the natural world. Instead of museums displaying rows of scientific specimens that served merely to instruct visitors on the number and form of species in the natural world, they could do much more. With lifelike taxidermied animals shown in their natural habitats and mounted and displayed with scientific accuracy, museum institutions could broaden the scope of their missions to include scientific education. This idea, made popular by Sir William Henry Flower, was already taking shape at the British Museum and spreading throughout Europe. The society would promote Flower's new museum idea in the United States by encouraging public displays with an educational purpose:

> Now, the mere fact that Museums *are* for popular instruction is a reason why the animals contained in them should be so arranged as to exhibit as many as possible of their most striking peculiarities and characteristics, and in order to do this some attitude in mounting must be permitted, and so far as is possible, an approach made to their natural surroundings. The Humming Bird should hover over a flower, the Woodpecker climb the side of a tree in search of food, and the Goatsucker should sit *lengthwise* of a bough suspended with outstretched wings and gaping mouth, as if in chase of insects. . . . In short, let each bird, so far as practicable, be mounted in an appropriate attitude and teach some fact in its life history.[25]

Lucas realized that natural history museums would fail to attract ongoing public support if their directors did not shift their focus to provide the audience with more than just "a general impression that there are a great many curious animals in the world."[26] For Lucas, such an impression failed to

communicate the true purpose of the museum. But make museum specimens attractive and give a sense "of their natural surroundings, varied attitudes, curious habits, food and mode of procuring it," and the audience
would "gradually gain some idea of its [the museum's] purposes, and appreciate the fact that it is something more than a mere collection of animals."[27]
Lucas noted that there were already two museums that had begun to implement this new educational model, the British Museum and the Princeton
College Museum. However, he also took the opportunity to chide Scott for
the first annual exhibition judges' decision to overlook "The Flamingo at
Home" by stating that although the Princeton museum "allows no turned
perches in the collection, and insists that the birds shall have as striking
attitudes as possible . . . many of the positions are odd."[28]

After a lively discussion of the topics, officers were elected for the following year: Lucas was elected president; Webster, vice president; Hornaday
remained secretary, and Fred T. Jencks was appointed treasurer. On the following evening, a large number of men and women attended a reception to
celebrate the public opening of the exhibition. There were 222 taxidermy
exhibits, as well as many articles of use and ornament. The *Boston Journal*
reported, "On every side there are curiously marked skins, rich furs and
gorgeous plumage, offering to the eye a variety, both of forms and of hues,
that is most attractive."[29] Although "A Fight in the Tree-Tops" and "The
Flamingo at Home" had already been judged at the first exhibition, they had
not been sold, and as a result were taken to Boston: "Facing the entrance
to the hall is a group entitled 'A Fight in the Tree-top' . . . [and] the place
of honor at the head of the hall is occupied by a group of Flamingos."[30]
The founding members felt strongly about the quality of Webster's flamingo
group and took every opportunity to buttress its reputation.

One of the new exhibits that apparently attracted a great deal of attention was an "Indian elephant two feet nine inches in height" and "not more
than six or eight months old when it came to its death."[31] This specimen was
mounted by Hornaday, but it was not a "baby" elephant, as the exhibition
catalogue described it. Most probably it was the fetus of the female elephant
that Hornaday had collected for Ward's in the Annamalai Forest in India.

Hornaday also contributed the only new habitat group featuring a mammal, titled "Coming to the Point," which received a specialty medal. It
featured a white setter dog "suddenly"[32] picking up the scent of six par-
tridges under cover of a dense bush of "autumn-tinted" leaves. The painted
background, by SAT member Mary E. W. Jeffrey,[33] gave the illusion that
the setter was stalking through a wooded area at the edge of a field on an
early morning in autumn. Hornaday had captured a moment in nature:

"Although the dog cannot see the game, his keen scent tells him it is very near, and he has come to 'a point' to indicate to his master the close proximity of the birds."[34] The group was mounted as a wall case, ten inches deep with a glass front surrounded by an ornate picture frame.

Hornaday considered the wall case of this type "one of the most popular and pleasing of all pieces of decorative taxidermy" and attributed its evolution to "the desire to protect from destruction the more cherished of the single specimens that first began to grace the homes of the lovers of animated nature."[35] For Hornaday, the case's wildlife painting was meant simply to enhance the taxidermy mount with a captivating motif that placed the animal in its natural setting:

> The tints of the picture should be very quiet, and by no means gaudy or striking, and should not attract attention away from the zoological specimens. The objects to be gained in a painted background are distance, airiness, and, above all, a knowledge of the country inhabited by the bird or mammal.[36]

Yet the society clearly valued the opinion of artists, electing James Carter Beard and Thomas H. Hinckley (along with naturalist J. W. P. Jencks) to judge the second exhibition. The judges did not favor "Coming to the Point," or any of the other larger groups, over the single animal specimens for the major awards. In fact, a solitary African monkey, titled "A Monkey Getting A Bite,"[37] mounted by Hornaday (whose baby Indian elephant was passed over), and a Caspian tern, mounted by Webster, shared the award for best piece in the exhibition. Despite overlooking the large mammal groups, the judges did favor many bird groups for the minor awards. Hornaday and the other SAT officers were satisfied with the judging, noting in *Ward's Natural Science Bulletin* that the judges' report "was received by the society with great satisfaction, and their criticisms will be remembered to good advantage."[38] Of special significance to the officers was that all of the entries demonstrated a general improvement in the quality of mounts over the previous year. The society was having a positive effect on its members, and the "highly meritorious"[39] quality of the taxidermy generated business for Ward's. The Boston Society of Natural History purchased Hornaday's "baby" elephant and placed an order "for a very fine black bear, a Canada lynx, fisher, fox and beaver, and other specimens are to follow."[40]

The society, however, did not fare as well financially. Expenditures for the second exhibition exceeded income. The most significant increase was in advertising.[41] Perhaps Ward's was not as well known in Boston as in Rochester, because although the society aggressively publicized the exhibition to ensure significant public attendance, ticket receipts increased by only $32.78 over the previous year. Hornaday blamed the Boston taxidermists for the meager attendance:

> With but four or five exceptions the sixteen professional taxidermists of Boston and vicinity treated the society with the utmost coldness and suspicion, and refused to identify themselves with the movement. In this respect they have shown themselves wholly different from all other taxidermists who have ever come in contact with our Exhibition or Committees, and their conduct was wholly without excuse or palliation.[42]

However, there were more tangible reasons why attendance was poor, including proximity to the Christmas holiday and a variety of events showing that same week: famed American actor Edwin Booth was giving "standing room only" performances in both *Hamlet* (the biggest box-office attraction of the nineteenth century) and *Othello* at the Park Theatre, while Gilbert and Sullivan's new comic opera *Patience; or Bunthorne's Bride*, "the Sensation of the Season," was showing at the Boston Museum, founded by Moses Kimball—a theatre ticket also purchased entrance to his cabinet of curiosities, best known for its wax figures. By charging a 25-cent entrance fee—the same amount as many of the theatrical shows (which cost between 25 and 75 cents)—the exhibition competed for the interest of the theatre crowd, a problem the society probably had not anticipated.[43] Whatever the reason, the society was in dire financial straits as it prepared for its third exhibition.

THIRD ANNUAL EXHIBITION

Only three months after the close of the second SAT exhibition, Ward lost two of his best men to the U.S. National Museum: Hornaday was appointed chief taxidermist in March, and Lucas began as osteologist in June. By this time, their friend and former Ward's co-worker, Charles H. Townsend—a member, though never an officer, of the SAT—was already working in Washington, D.C., as a field naturalist for the U.S. Fish Commission. Their paths would cross again many times over the next three decades as their three separate, but parallel, careers took them from Washington to New York City. Webster would soon follow the group to Washington, although he chose to establish his own private taxidermy studio there.

George Brown Goode, the newly appointed curator of the National Museum, along with its secretary, Spencer F. Baird, had been following the SAT from its inception. (Goode and Baird had been named honorary members in 1880.) Goode, in particular, was keen to hire the society's originators, as he attributed to them the founding of the "new American school" of taxidermy. He believed that the Smithsonian was the best environment for the society's founders to work out "the ideals of the organization." In retrospect, Goode said that the museum would have hired others as well "but for our feeling of unwillingness to interfere with the important establishment at Rochester by taking away so many of its most competent men."[44] Goode held Henry A. Ward in high esteem, believing that his "hope of profit" was subordinate to his "love of natural history and the ambition to supply good material to museums"—ideals he considered "not very usual in commercial establishments."[45]

Even though the museum's administration was supportive of the SAT, the new preparators' first responsibility was to design exhibits and mount specimens for the museum. It was therefore difficult for Hornaday and Lucas to find time to organize the third annual SAT exhibition. Although Lucas was on hand in Washington, he had always had a lukewarm interest in the society and served more as a self-described "thorn in the spirit."[46] The bulk of the responsibility for advancing the SAT thus fell to Hornaday, who remained committed to the society and, more importantly, to its mission.

As he adjusted to his new duties at the Smithsonian, which must have consumed a considerable amount of his time, Hornaday found it increasingly difficult to organize key SAT members through correspondence. Yet this was neither Hornaday's nor the SAT's greatest challenge. The financial loss in Boston, which had to be absorbed by a few of the society's members—including Lucas, W. E. D. Scott, Jules Bailly, and Thomas Fraine, among others—left the society's bank account empty. Hornaday thus began to look outside the SAT for funds sufficient to secure at least a building in New York to house the next exhibition.

By October 1882, Hornaday had found a patron: Jacob H. Studer, a publisher of natural history books who had just published the fourth edition of *Studer's Popular Ornithology: The Birds of North America: Drawn and Colored from Life*, donated a copy of the book to be auctioned off[47] to help the society pay a bill owed to the *Rochester Democrat and Chronicle* for services rendered at the first exhibition. More importantly, he offered to "advance" the society the sum of $500 for the third annual exhibition. Studer urged Ward to match half this sum with an additional $250.[48] Only if the exhibition proved to be a loss, as in Boston, would the money become a donation. Hornaday's efforts had caused the society, in his own words, to "rise like a giant refreshed." In response to Studer's generosity, the officers decided to create a board of exhibition commissioners to assist in the planning of the exhibition, allowing the taxidermists to devote more time to preparing specimens. Studer was elected president; George Brown Goode, vice president; Dr. J. B. Holder, secretary; Andrew Carnegie, treasurer; and Dr. Wendell Prime, James C. Beard, Henry A. Ward, Robert Colgate, and Allen S. Bickmore were chosen to complete the commission. As president, Studer took charge of organizing the exhibition. Hornaday's unwillingness to relinquish control to Studer quickly incited a contentious relationship. Ward was the obvious referee between the two. That fall, he and Hornaday exchanged a feverish correspondence in an effort to smooth ruffled feathers.

The major point of contention was that Studer had promised the loan for an exhibition that was to be held in December, but by late October there was already talk of postponing it until February of the following year. The taxidermists contributing exhibits had urged Hornaday to delay it, and he, too, saw the benefits of delaying the show. The failure of the Boston exhibition loomed large in their minds, and they were all reluctant to repeat the mistakes of the past; more time would at least ensure a larger number of exhibits. Hornaday may have felt overwhelmed by his new position as the National Museum's chief taxidermist and saw the delay as an opportunity to balance his commitments. Goode helped to ease the burden of preparing taxidermy mounts for the exhibition by encouraging both Hornaday and Lucas to enter the specimens they were already preparing for the museum. However, Hornaday was uncertain that he would enter anything at all, as he believed there was little reason to compete now that he had achieved the highest-ranking position for an American taxidermist. Even so, he proved that the competitive spirit was still a great motivation for him:

> I do not intend, unless I have good reason to change my mind from what it is at present, to enter anything in competition this year, and perhaps never again, unless I am forced to it by insinuations that somebody else can lay over me on mammals. This would render a trial of skill necessary. Now that I am here, I have no further desire to enter into any rivalry with my old colleagues, so long as my position is fairly conceded. But and until people begin to indulge in odious comparisons it won't be necessary. Lucas informed me the other day with a grand flourish of triumph that Baker had just written him that "the mammals they had just shipped to Central Park were better mounted than any that had ever gone to an S.A.T. exhibition, or anywhere else." Now *that* was for my especial benefit, free of charge and no drawbacks on account of shrinkage. And it had better not occur again unless your boys desire that I should

try conclusions with them some more. If it was *true*, then I must enter the lists as usual against all comers. If it was *not* true it ought not have been said. If there is any further doubt in the mind of anybody as to my ability to take the lead on mammals *both large and small*, of all kinds, why then I want to see that question definitely settled, and settled it shall be.[49]

One week after confiding to Ward his rancor toward the other members of the society, Hornaday decided to enter several items into the competition: a polar bear, a cinnamon bear, a seal, an antelope, four mammal heads, an ibis case, and one rug.[50] More importantly, he had used a new taxidermal method to mount an elephant and a Chihuahua, which he entered as most likely to demonstrate his expertise in both large and small mammals.

Hornaday perceived threats from all sides, with the exception of Henry Ward, with whom he continued to keep up a weekly correspondence. By December, his relationship with Studer had deteriorated into name-calling:

> Studer has got his back up about the postponement, in fact he went back on the *whole business* as soon as I was out of N.Y. and now says, "Prof Ward and yourself can come on here and run the Exhibition to suit yourselves." . . . I am *disgusted* with him. Now only one thing remains, and that is for *you* to become President of the Board and let him go to the d—l, where he belongs.[51]

One week later, Hornaday still raged at Studer: "It will be war to the knife as far as my strength will carry me. He shall have the hostility of every member of the S.A.T. and many others of greater influence."[52]

By late December, the new April exhibition date was set. It was announced in the January 1883 issue of *Ward's Natural Science Bulletin* in a column about the society, in which Studer's loan was celebrated:

> Members and friends . . . have reason to rejoice in the fact that its prospects, for the immediate future at all events, are so bright and promising. . . . A recent favorable turn in the fortunes of the society has caused it to "rise like a giant refreshed." . . . Mr. Jacob H. Studer. . . . generously offered to advance $500 as a guarantee fund for the expenses of the exhibition.[53]

Ward also noted that Studer had been elected president of the SAT Board of Exhibition Commissioners. There is little doubt that he hoped the whole mess would blow over; yet clearly Ward was working on behalf of the so-ciety, as the announcement of the loan in print would make it much more difficult for Studer to go back on his word. The *Bulletin* also included Hor-naday's favorable review of *Studer's Birds of North America*, even though Hornaday had vowed to "injure . . . his book" if Studer reneged.[54] Despite Ward's efforts, in the end there was no resolution, and Studer withdrew his loan and resigned from the board. Dr. Joseph B. Holder, not Ward, was elected the board's new president.

Fortunately for the society, Hornaday, while on a collecting trip for Ward's in 1879, had met Andrew Carnegie in Singapore when they dined at the American consul's house. Carnegie, at the time, was traveling around the world with his assistant, John W. Vandevorst. He was immediately taken with Hornaday's stories of collecting. Hornaday recalled that the "idea that orang-utans, dugongs and great snakes had a market value and could fluctuate was to him the funniest thing yet found in the Far East."[55] In his book *Round the World*, Carnegie humorously wrote of the encounter:

> The recital of his adventures are extremely interesting. . . . In the ab-sence of other commercial intelligence, I may quote the market in his line. Tigers are still reported "lively," orang-utans "looking up"; pythons show but little animation at this season of the year; proboscis monkeys on the other hand continue scarce. There is quite a "run" on lions, and kangaroos are jumped at with avidity. Elephants are "heavy"; birds-of-paradise drooping; crocodiles are snapped up as offered, while dugongs bring large prices. What is pig metal to this?[56]

Upon returning home, the two men began a lifelong friendship. As Carne-gie's philanthropic interests turned toward diverse social and educational institutions, he was naturally interested in the endeavors of the SAT. He agreed to serve as the society's treasurer for the third annual exhibition. Once Hornaday was certain that Studer would not honor his pledge, he went directly to Carnegie, who without hesitation gave the society the en-tire sum and an additional monetary gift that erased all of the debt that it had accrued in Boston. To show the society's appreciation, the officers named Carnegie an honorary member and presented him with a peacock fire screen and a white heron medallion as souvenirs of the exhibition.[57]

On Monday, April 30, the third annual exhibition—held at Lyric Hall, on Sixth Avenue between Forty-First and Forty-Second Streets, in New York City—opened with a by-invitation-only reception. The press corps was also invited to preview the exhibition. Ward began the evening's event with a brief explanation of the society's purpose, which he defined as motivated by a "self-culture."[58] He then introduced Joseph B. Holder, curator of invertebrate zoology at the AMNH, who delivered the honorary address.

From the perspective of an experienced museum curator and science educator, Holder spoke of American taxidermy's past, present, and future. He defined the "museum of old" as "one that did not always encourage the best art," and announced that the new museum, the "scientific museum," "which calls for the best and a great deal of it," had taken its place.[59] He credited the geological and topographical surveys of the American West with creating a need for naturalists skilled in taxidermy, referring to Audubon's tour of the western plains, which would not have been so successful without John Graham Bell, one of America's first notable taxidermists, who also was present that evening. He credited the Smithsonian Institution with being "a sort of patron of the art on a large scale, both through the publication of directions for the preservation of specimens, and by furnishing ways and means by which parties going out could successfully explore the regions likely to furnish desired material. . . . The vast storehouses of our museums now attest the advantages accrued therefrom; and this is the work of a few short years."[60] While he praised museums for making strides in employing taxidermists skilled in preparing specimens, he noted that the art in the recent past was considered a "mere cypher," but that now the new taxidermy— "moulding the skins of quadrupeds and birds and fishes and snakes to resemble life"—was recognized as an art of "great capabilities."[61] He believed that Jules Verreaux, the celebrated pioneer of naturalistic taxidermy, had obviously influenced the new American school, but he believed that the nation's taxidermists could also look to America's zoological illustrators, such as Audubon, to inspire lifelike taxidermy mounts. He then challenged the new taxidermist to "exercise all his best faculties" in creating a superior art, calling on him to learn to observe nature, sketch, and model:

> There would seem to be much in the modelling in clay or other material that would teach the eye to catch the requisite form readily. Indeed all appliances that aid in the advancement of the artistic faculties. . . . Close observation of nature, of living natural forms, their characteristic attitudes under the various conditions likely to occur, all are of infinite service.[62]

To illustrate his meaning, he harked back to Webster's ever-controversial flamingo group, finally revealing the reason why the judges from the first exhibition did not choose the group:

One of the feet, which in some species of birds having the legs extended, as in this case, would naturally bend abruptly in the instep, was placed flatwise upon the ground, and this was judged to be a defect. A reference to the attitude of the living bird showed the artist to have been a close observer; he was correct. An instantaneous photograph of a similar bird plainly indicates this feature. . . . This may teach us to look more closely at living examples.[63]

This public announcement of the judges' error was more than just another opportunity to praise Webster's flamingo group; it was a recognition that the society's taxidermists were of a class on par with the scientists who curated the nation's natural history collections, and a further endorsement of the new American taxidermy movement.

Holder emphasized the importance of the SAT's mission by concluding his lecture with a brief discussion of how the relationship between the art of taxidermy and natural history museums had become one "of the greatest importance." No longer were taxidermists creating, in his estimation, "stuffed horrors, too absurdly prepared for any one's pleasure." Recent advances in the art had made it possible for the nation's museums to proudly display their "pet pieces of taxidermy." However, he conceded that a dearth of good taxidermy was not entirely to blame for the failure of old natural history museums to hold the interest of the public, referring to the fact that although the AMNH had been founded nearly fifteen years earlier with a mission that proposed to wed popular instruction and the natural sciences, the experiment had thus far been a near failure. But Holder believed in that dual mission, and he saw the new taxidermy as a means by which to garner public support: "Good art . . . is a mighty power, and we now begin to see how influential it is."[64] As the author of *Elementary Zoology*, published in Appleton's natural science series for high-school-aged readers, Holder understood the changes that needed to take place for natural history museums to present valuable public instruction, and as a senior scientist at the AMNH, he had the influence to win the support of other curators.

Lucas followed with a paper on "The Scope and Needs of Taxidermy." He concurred with Holder's estimation that the new taxidermy would transform the old museum, where "already there is a faint rustle among the dry bones and lifeless skins . . . the day is coming when we will witness a revolution in the style of our museums." He also credited American museum directors with recognizing, first, that museums are "largely dependent on taxidermists for their most attractive and instructive features; and, second, that there is a difference in the quality of workmanship." However, he insisted that the reason museums lacked quality specimens was not that taxidermists were only now beginning to create quality mounts, but rather that museums of the past were not willing to pay for quality "exhibit" specimens. Museums like the AMNH had come to realize that "low-priced work is not always the cheapest in the end."[65]

Sensing a long-awaited sea change, Lucas believed museum taxidermists could turn their attention to the "scope and needs" of the new movement. The idea that the scope of taxidermy would now include its reclassification as a fine art was not unheard of in this period. In fact, eighteen years earlier, Frederick Law Olmsted, founder of American landscape architecture, together with his co-designer, architect Calvert Vaux, in developing the profession of landscape architecture, argued that their work, then considered an extension of the natural sciences, should be raised to the level of the fine arts. Vaux and Olmsted together saw that this emphasis would help to promote their cause and garner public support.[66]

Like Vaux, the majority of taxidermists were passionate about their inclusion in the fine arts. Lucas thus explained that "as the artist makes it possible for us to see the beauties and grandeur of a landscape that we can never hope to behold ourselves, so it is on the craft of the taxidermist that we must rely for ideas of most animals, and on the amount of his skill depends the correctness of our impressions."[67] He emphasized the urgency of the taxidermist's task: "Our wild animals and especially the larger ones, are being rapidly civilized from the face of the earth," so museum specimens would eventually be all that remained of some species.[68] Taxidermy, he argued, was more accurate than drawings and paintings, because "none but the very best of paintings produce anything like the impression aroused by the animals themselves. We may admire a painted Tiger, but we feel no dread of him, and while we may realize from the figure of an Elephant that he is large, we still fail to fully appreciate his true size."[69] Museums, Lucas would later write, "have an enviable reputation for the manner in which they hold the mirror up to Nature,"[70] but he clearly believed it was the taxidermist who held the ability to reflect realistic images of nature for the edification of the museum visitor. Unfortunately, Lucas lamented, the state of the field was such that "too often our finished specimens are creatures of our imagination."[71] He then went on to describe the guidelines by which taxidermists across America could begin to reform the field of taxidermy—specifically, by careful study of the forms of animals, which would improve greatly the taxidermist's artistic and mechanical principles. Then taxidermists could begin to establish acceptable practices for mounting specimens. Finally, he urged taxidermists to compare their work, "freely admitting that our own is perhaps not perfect, and striving to profit by the excellencies of our neighbor." He went on to insist, "We must get rid entirely of the idea that but one man is master of the art, and that one man is ourself." This statement was a direct nod in Hornaday's direction. Because Lucas had never aspired to being the *best* taxidermist, he, more than any of his contemporaries, recognized that the competitive nature of the art could lead to its demise if it was not harnessed productively.

The third annual exhibition displayed both the best and the worst of what Lucas had described. The unfortunate divide between high- and low-quality mounts was marked. As *Forest and Stream* complained:

> The visitor enters the hall, expecting to find all the work of a very high order of merit, and instead of this he sees amid much that is good, a great deal that is commonplace, and more or less that is positively bad, and unworthy of a boy who has not yet mounted a hundred specimens. . . . The society this year exhibits with many excellent productions, much that is very wretched. It could scarcely be otherwise, for all that is sent in for exhibition must be accepted, or else jealousies and heartburnings would arise, and the exhibitors whose pieces were rejected would feel that they had been badly treated by the society.[72]

Although the article misses the point of having an annual exhibition for the purpose of the edification of all taxidermists, it does indicate that there was public expectation and demand for quality museum mounts.

Once again, Hornaday's entries caught the attention of the public as well as the judges. Given that he entered specimens at this exhibition to settle the question of who was the country's best taxidermist in both large and small mammals, it is not surprising that he entered not only an African elephant, Mungo, but also, as a comparison, a Chihuahua—referred to as a "hairless Mexican terrier." The two mounts served to demonstrate the success of what Hornaday later termed the "clay-covered hollow-statue" method for mounting large mammals. While Martha Maxwell used this method before Hornaday, she did not publish an account of it, and so Hornaday was free to claim that he was the first American taxidermist to use clay as a filling material. In fact, in *Taxidermy and Zoological Collecting*, Hornaday wrote that "previous to 1880 its use among the taxidermists of my acquaintance was unknown, and when its value was discovered and put to general use by the writer . . . many of my rivals predicted all manner of evil prognostications, and now its general use really marks a new era in American Taxidermy."[73]

Hornaday hoped to show that the clay method allowed attractive mounts even for mammals that were nearly hairless, from the largest to the small-est.[74] At one extreme was Mungo, a six-year-old African elephant that had died in Washington while working in Adam Forepaugh's circus and menag-erie, the greatest rival of P. T. Barnum's circus.[75] Forepaugh's manager sold the specimen to the National Museum. Hornaday, in describing the elephant to Ward, referred to Mungo as "African, but not Jumbo"—he was five feet tall, half the size of Barnum's great elephant. A specimen of this size was perfect for Hornaday's experiment with his new method. Mungo was large enough to demonstrate the clay method's effectiveness, but not so large that Hornaday could not overcome problems due to the weight of the skin or the difficulty of modeling such quantities of clay.

Even at his relatively small size, Mungo presented numerous challenges. The importance of taking careful measurements of the living animal when possible, or the recently deceased one, stressed often by the society, was particularly significant in the mounting of mammals with little hair. Thus Hornaday began the process of mounting Mungo by measuring "the body, showing its length, height, and girth at various points, and the dimensions of the limbs and the trunk. These were supplemented by sundry drawings, and by plaster casts of the head and of the limbs of one side." Another ob-stacle to overcome was that the skeleton, generally used inside the mount, was to be mounted separately. Hornaday therefore began the process of cre-ating a "false body," or manikin:

> The backbone . . . consisted of a broad two-inch plank, the upper edge of which was carefully cut into an exact copy of that dorsal outline which is so characteristic of the African elephant. To this the legs were at-tached by heavy angle-irons, the iron that formed the axis of the leg running through a hole in the free arm of the L. The legs themselves were formed of excelsior solidly wound around roughly hewn wooden bones. . . . The broad overhanging pelvis was next added; and then the skull, with its massive jaw, was built on, the more salient portions being carved with care from the plaster model, while those buried deeply in the flesh were more roughly copied.
>
> The long ribs of the original were represented by bands of iron wrapped in tow, fastened above to the plank backbone, and below to the underside. A neck of laths, covered with excelsior, joined the head to the body. Wooden shoulder-blades were now put in place, the tail and trunk added, and then, following the diagrams and accompanying measure-

ments, the vacancies existing between the upper parts of the legs and adjacent portions of the body were carefully filled out.

Lucas, in an article for *Science*, with admiration likened this stage of the process to the magical creation of Feathertop, the scarecrow brought to life and transformed into a gentleman by Nathaniel Hawthorn's character Mother Rigby: "The elephant at this stage stood forth a creature of wood and tow, only waiting for the final metamorphosis which should fill the blank wooden orbits with twinkling eyes, and endow the entire framework with the semblance of life."

Before Mungo could be brought to life, the skin had to be placed over the frame. The skin was first secured along the back, and its underside was then covered with a "thin coating of clay mixed with chopped tow." Working quickly, one taxidermist applied the clay mixture while the other positioned the skin and sewed it together—first the midsection, and then "one by one the legs, trunk, and tail were similarly treated, the skin being covered each night with wet cloths to preserve it moist and flexible throughout." When the sewing was complete, the fine work of adding wrinkles to the skin was accomplished with a pointed modeling tool. The more difficult skin folds, particularly those of "the trunk, elbows, and flanks, were secured by wires or twine to hold them in place until dry." Once the animal's face was inscribed with its final expression, Mungo's glass eyes, "made from a color-sketch of the originals," were inserted. When the mount had thoroughly dried, the seams were "filled with *papier-maché*" and "a slight but careful use of color restored the skin to its original aspect." The new clay method thus solved the problem of mounting a lifelike elephant with its hide naturally wrinkled "instead of, as is too often the case, smooth and swollen."

Contrary to Hornaday's belief that even those close to him were intent on proving that they could outdo him in taxidermy, Lucas proved his esteem for Hornaday and his work when he concluded the *Science* article by stating that Mungo represented "the beginning of the new and better class of taxidermy at the national museum." *Scientific American* described Hornaday's elephant as "the most perfect work in the exhibition" and "as a specimen of the taxidermist's art [that] is superior to anything yet done, in this country at least . . . and it does not seem as if the original flesh and blood Mungo could have been more lifelike. Mr. Hornaday . . . has shown himself an artist as well as a taxidermist."[76] Forty years later, Hornaday reflected on Mungo with characteristic conceit: "If we had him to do all over again today, we could not improve upon the original edition, and we suspect that it

Fig. 2.3. William T. Hornaday's mount of the African elephant "Mungo" wrapped in tow. The mount was an experiment in Hornaday's "clay-covered hollow-statue" method for mounting large mammals. (Smithsonian Institution Archives. Image #MNH-2791)

is not every 'sculptor-taxidermist' who is destined to view with smug complacency his work on large mammals forty years after its perpetration."[77] Although Hornaday's clay method was adopted by other taxidermists and improved upon only a few years later, Mungo stood as a triumph in the new art of taxidermy, and Hornaday certainly was the American father (and

Martha Maxwell the mother) of the clay model method for mounting mammals. Even Mungo's tiny counterpart, the hairless Mexican terrier, was a wonder to all of the exhibition-goers. *Scientific American* reported that the dog "was equal in merit to the elephant" and that it "was passed over by the judges in consequence of an impression that it was a plaster cast. Certainly

Fig. 2.4. William T. Hornaday's mount of the African elephant "Mungo" wrapped in tow and draped with skin. The mount was an experiment in Hornaday's "clay-covered hollow-statue" method for mounting large mammals. (Smithsonian Institution Archives. Image #MNH-2789)

Fig. 2.5. The African elephant "Mungo" in completed form, as displayed at the Third Annual Exhibition of the Society of American Taxidermists in New York, 1883. (Smithsonian Institution Archives. Image #MNH-2788)

a tribute to the excellence of the work."[78] R. W. Shufeldt, nearly ten years later, described the mount:

> This dog had no hair at all apparently, and his skin was as thin as ordinary writing paper, but through the aid of a plaster cast of his entire body as a model and the use of the clay-covered manikin, a most remarkably fine thing has been produced. This specimen has also been delicately tinted where it became necessary, and as now preserved will last without change for an indefinite length of years.[79]

Ward's Natural Science Bulletin argued that only a taxidermist could "properly appreciate the difficulties to overcome" in mounting this animal, finding that "the only drawback of this piece lay in the coloring of the skin, and we frankly confess ourselves unable to remedy that. The difficulty lies in applying paint so that it will appear to have some *depth* to it and allow the texture of the skin to show as in life."[80]

Though the Chihuahua was the most popular of the dog mounts, the exhibition also featured Hornaday's "Coming to the Point" from the Boston exhibition, which was hung on the wall to the right of the entrance. Hornaday's style of mounting a dog in a wall case immediately found a market among bird hunters who wanted a permanent monument to their favorite companion. Two other dog mounts, hoping to capitalize on this interest, were also on display, but *Forest and Stream*—a favorite magazine among the potential clientele—complained that not one of the three mounts quite looked like a live dog. "Coming to the Point" was described as having an excellent background and accessories, but the mount looked like no dog the writer had ever seen. "He is long and thin, and has a very small head, a minute head in fact," he wrote. "Evidently the skin has been very much stretched. The head is very fair, but the rest of the body is all out of proportion to it."[81] Across the aisle was Thomas Fraine's mount of H. H. Warner's "Old Frank," described as a "brown and white pointer, standing a pair of ruffed grouse." The writer preferred Fraine's overall design to Hornaday's, claiming that "the case, the ground and the birds are even better than with the setter." However, he was more critical in regard to the body of the dog: "It is round and without shape; the body of a dog, in fact, which is so fat that not a bone is visible. The flanks are not drawn in, not a rib nor a vertebral spine can be seen. A dog shaped like this could not and would not hunt an hour." Two weeks later, *Forest and Stream* printed Fraine's explanation of the awkward mount:

> Old Frank was eleven years old, and very fat, and furthermore was sick for a year with a tumor. Mr. Warner tried all the known medical skill to cure him, but to no avail, hence he was turned over to my hands to chloroform. His fatness was no fault of mine, and previous to his sickness he was as good and staunch a dog in the field as ever stood on four legs.[82]

Forest and Stream, however, did give partial approval to at least one of the dogs, selecting John Wallace's black-and-white pointer on a covey of quail as "the best stuffed dog on exhibition. The attitude is not nearly so well chosen as that of the other pointer, but it is all over a dog, and were it as well shown as the other two it would far surpass them."[83] For the first time, an exhibitor other than a founding member of the SAT received public praise for his taxidermy. Clearly, the overall quality of the exhibition work was improving, and the society was beginning to attract membership from more established taxidermists like Wallace, who saw the society as a way of promoting his more intricate and expensive work.

Wallace was a well-known taxidermist in New York City, but his workshop was modest—a tiny basement on North William Street under the Brooklyn Bridge that, Lucas wrote, could "by no stretch of the imagination . . . be called a studio."[84] Still, his shop had earlier been frequented by the likes of Spencer F. Baird and O. C. Marsh, and young naturalists such as C. Hart Merriam apprenticed there.[85] Though Wallace produced mounts for the U.S. National Museum and the AMNH, and even for a time ran his own museum in his hometown of Paterson, New Jersey, Lucas disapproved of Wallace's work, viewing it as representative of the old style of taxidermy.[86] He claimed that Wallace "probably stuffed, most literally, more animals than any other one man" and condemned his business as merely "a commercial establishment, and particularly one that dealt mainly with the preparation of single specimens for museums."[87]

However, Wallace's success may have been the result of the work of his young assistant, Carl E. Akeley. Akeley had been employed at Ward's—one of those who had filled the vacancies left by Hornaday and Lucas in 1883—but he had been fired after a feud with Ward.[88] By the time the third annual report of the SAT was issued, Ward had apologized to Akeley and hired him back, but leading up to and during the third annual exhibition, Akeley was working for Wallace—and probably assisting or working alone on the mounts entered under Wallace's name. Akeley's skill might help explain why Wallace, who was considered a hack practitioner of the art, received numerous awards at the third exhibition, including the silver medal for the best exhibit of heads; specialty medals for "Tartar Hunter Attacked by Lions" (an homage to Verreaux's "Arab Courier Attacked by Lions") and "Great Horned Owl at Bay"; the diplomas of honor in "Taxidermy Proper" for the pointer dog and quail, a lioness, male and female albino deer, a monkey, and a bald eagle; very high commendations in mammals for both "Lions Fighting over Their Prey" and a single monkey specimen for composition and dramatic effect, and in birds for a display of owls; and high commendation for the bald eagle mount. Wallace's dramatic pieces were so successful in drawing visitors to Lyric Hall that during the week of the exhibition, *Harper's Weekly* ran a half-page engraving by Daniel Beard, one of the judges, that featured "Lions Fighting over Their Prey" as its centerpiece.[89]

Though they were far less dramatic than Wallace's exhibits, among the habitat groups, *Forest and Stream* judged that "two of the most strikingly beautiful are a group of duck bills . . . and one of terrapins." Webster mounted the duck-billed platypus group, for which he received a commendation "for evident accuracy and study."[90] The group, displayed in a case, included a male, female, and young. While Ward was on a collecting trip in Australia, he observed platypus behavior and collected nine specimens of different sizes and ages. He then conceived of a group display that would demonstrate the animal's life history, which he intended for Webster to mount. The result was groundbreaking:

> The group represents the strange little creatures at their home in a bend of the river bank, where a clear shallow pool with lilies on its surface and sedges at its sides allows them to dispose and place themselves according to their natural habits, coming from and entering their burrows, swimming in the water, digging in edge of bank, rolled up in sleeping attitude, lying prone to sun themselves and disporting on a limb of tree over-hanging the water. It also shows the vegetation of the locality in which they live as closely as possible . . . a caving of the bank has disclosed the interior of a burrow and the narrow subterranean passage . . . which is below the surface of the water.[91]

It was noted that "the air of activity and of domestic cheerfulness" made the exhibit "peculiarly pleasing," and thus "attracted great attention."[92] The AMNH purchased the exhibit for $325 and installed it in its Hall of Mammals.[93] Despite Webster's achievement, it was Hornaday's African elephant that received the silver specialty medal for the best piece in the entire exhibition and Webster's "Wounded Heron," not the platypus group, that received the bronze specialty medal for the second-best piece. The bird, actually a white egret, was "transfixed with a golden arrow," with blood dripping from the wound, and "mounted against a blue velvet background."[94] *Ward's Natural Science Bulletin* reported that this beautiful mount was "perhaps the finest example of thoroughly artistic taxidermy we have ever seen."[95] Although the critique was leveled that "the left wing was in an attitude unattainable by the living bird," in Webster's defense it was noted that such a position might be attainable in a struggling bird.[96]

The other group praised by *Forest and Stream* was Lucas's "edible terrapins," for which he received the bronze specialty medal for the second-best exhibit of reptiles. The group represented four species and three families of North American chelonians (a diamondback terrapin, a yellow-bellied terrapin, a red-bellied terrapin, and a snapping turtle) in a single habitat. The specimens were placed above and below the "water." Only one is in the act of diving (probably the snapping turtle—the only species in the group with

webbed feet), its body above and below the water—in Hornaday's estimation "a successful accomplishment of a very neat mechanical feat."[97] Lucas
intended the group to reveal the possibilities that existed in mounting "animals, which like turtles do not readily lend themselves to the making of
groups. Birds and mammals can be much more easily combined in artistic
and instructive groups."[98] Eight years later, Hornaday praised the arrangement in *Taxidermy* as being an "altogether unique and pretty group [that]
teaches one very important lesson, viz., that even the most commonplace
animals are interesting when they are well mounted, and grouped with a
setting which represents their natural haunts."[99]

The most unusual exhibit, "A Taxidermist's Sanctum: The Proprietor at
Work," located in the south room of Lyric Hall, was the combined effort of
the taxidermists of the U.S. National Museum. In April 1882, one month
after Hornaday was appointed chief taxidermist, Secretary of the Smithsonian Spencer F. Baird commissioned the society to prepare "a collection
of objects illustrating the present condition and possibilities of the art of
taxidermy" that would not only serve in connection with "the other educational features of the museum," but would also "call attention to the avail-
abilities of taxidermy in various branches of the decorative arts, and . . .
stimulate competition among taxidermists, and thus encourage a higher
degree of excellence in workmanship."[100] "A Taxidermist's Sanctum" was
the first installment of the museum's taxidermy exhibit. The figure of the
taxidermist, designed by John W. Hendley, was seated at a workbench on
which are placed the several tools and materials used in the mounting process. He was surrounded by specimens in various stages of preparation as he
bent over the bench preparing a bird. The *New York Commercial Advertiser*
printed a humorous account of the exhibit, claiming that "so life-like is the
figure that a gentleman asked several questions of it yesterday regarding
the process."[101] The exhibit was intended for the visitor who had no understanding of the art and science of taxidermy and who viewed the taxidermist's workroom as "in itself a curiosity shop."[102]

After the exhibition, "A Taxidermist's Sanctum" was removed to the
U.S. National Museum, along with several of the award-winning mounts
from the three annual exhibitions, including Hornaday's "Coming to the
Point," Lucas's "An Interrupted Dinner," Webster's "Wounded Heron," and
John Wallace's bald eagle. The SAT believed that this exhibit would be the
"first official recognition of taxidermy as a fine art" and become "a lasting
monument to the society." When the third annual report finally appeared in
the early summer of 1884, Hornaday, as the new SAT president, highlighted
the exhibit by including a full-plate photograph, and in his brief "Purpose

Fig. 2.6. Contributions from the New York exhibition of 1883 to the taxidermic collections at the U.S. National Museum, as photographed for the third annual report of the SAT. (1) "Coming to the Point" by Wm. T. Hornaday; (2) "An Interrupted Dinner" by Frederic A. Lucas; (3) head of caribou by J. Wm. Critchley; (4) peacock screen by Thos. W. Fraine; (5) "Wounded Heron" by Frederic S. Webster; (6) dead gull by Elwin A. Capen; (7) great horned owl by John Wallace; (8) bald eagle by John Wallace; (9) fox squirrel by P. W. Aldrich; (10) hummingbirds by Mr. and Mrs. G. H. Hedley; (11) "Nutcrackers"—squirrels by Joseph Palmer; (12) south-southerly ducks by Wm. Palmer; (13) "Sold Again" by J. F. D. Bailly; (14) frogs, toads by J. F. D. Bailly; (15) snowy egret by Thomas Rowland; (16) portrait of Jules Verreaux presented by J. F. D. Bailly. (Smithsonian Institution Archives. Image #MNH-2783)

of the Society," pointed out what he considered to be its most significant achievement:

> One great desire of the members is to raise the standard of museum work, so that American museums may lead the world in the quality of their material and be filled with lifelike animals instead of being storehouses of monstrosities. It is gratifying to know that nearly all the work now done for our large museums is done by members of the society.[103]

FOURTH ANNUAL EXHIBITION

Whatever degree of victory may have been celebrated after the third annual exhibition, it was to be short-lived. On July 30, 1884, the SAT held what would prove to be its final meeting in the lecture hall of the U.S. National Museum.[104] The members gathered to discuss Goode's invitation to exhibit at the World's Industrial and Cotton Centennial Exposition in New Orleans—celebrating the hundredth anniversary of the production, manufacture, and commerce of cotton in the United States—and thus defer until the following year the "regular" fourth annual SAT exhibition, which was to be held again in New York City. The society would share space with the National Museum exhibit in the Government and State Building, and Goode offered to pay all freight charges and the expenses to send a committee of SAT members who would supervise "the reception, handling and arrangement of the display, and who would look after the interests of the society in general and the exhibitors in particular." The group unanimously accepted the invitation. In keeping with the format of the SAT exhibitions, Goode contacted E. A. Burke, director general of the New Orleans Exposition (an influential southerner who was editor of the *New Orleans Times-Democrat* and treasurer of Louisiana), and suggested that the exposition sponsor a taxidermy competition. Burke agreed and offered to award gold medals for the best group of birds, the best single piece, and the best general exhibit, as well as second-place diplomas.[105] However, he reserved the right to withhold the awards in the event that the best entry in any category was not found worthy of high honors.[106] Once the details were worked out, Hornaday, in a rushed circular to the society, emphasized the importance of the exhibit and the patronage of the U.S. National Museum:

> It is hardly necessary to call attention of the members of the importance of the display to the profession generally, or to the efforts that the officers of the National Museum are making to advance the interests of the society. At this Exposition the organization will enter a new field, and it is reasonable to suppose that if the present opportunity is properly improved the results cannot be otherwise than advantageous.[107]

Unfortunately, Hornaday was alone in his enthusiasm, as only he and Lucas, probably grudgingly, formed the "exhibit committee."

The final SAT display included "A Taxidermist's Sanctum" from the New York exhibition and award-winning mounts from society members that Hornaday had been gathering to form the permanent SAT exhibit for the U.S. National Museum. Despite hurried preparations, the Smithsonian exhibit reports boasted that the SAT exhibit "contained specimens of the best work of the leading members of the society, including Messrs. Hornaday, Lucas, Fraine, Webster, William Palmer, Joseph Palmer, Hedley, Forney,

Bailly, Wallace," and others.[108] The popularity of taxidermy and the SAT is evidenced by its recognition in the press. The *National Republican* and the *Washington Post* both ran two-column feature articles promoting the New Orleans Exposition, and each included a mention of the SAT exhibit.[109] Near the close of the exposition in 1885, the *New Orleans Times-Democrat* praised the SAT, writing that "the society has accomplished its purpose of proving taxidermy to be one of the fine arts," and credited Hornaday and his taxidermy with its success.[110] Ironically, after *Forest and Stream* had so criticized "Coming to the Point" in New York, this time its correspondent described the piece as "beautiful beyond description" and commended the SAT's entire exhibit for its "beauty of execution and artistic design."[111]

The New Orleans Exposition ended in May 1885, and in June Hornaday supervised the return of all the exhibits to Washington. That same month, he installed the SAT exhibit at the National Museum. There never was a fourth annual exhibition. Without Hornaday—who had turned his attention to the West, where the potential extinction of the American bison meant that specimens needed to be obtained for the National Museum's collection—the society lost all momentum. No formal dissolution appears to have occurred. Hornaday later argued that the SAT had accomplished its goals and was no longer needed, while Lucas believed that it was too ambitious and could not be sustained.[112]

To some extent, both men were correct. From Lucas's perspective, it was too difficult to organize all taxidermists under one umbrella: even at the height of the society's popularity, it only managed to attract a little over one hundred members, all from the East Coast, particularly Boston, New York, and Washington, even though there were hundreds of practicing taxidermists across the country. While it was advantageous for museum taxidermists to share their knowledge, commercial taxidermists benefited from keeping their methods secret. They would always be interested in receiving awards for their work, but they would never participate in sharing their expertise publicly. Nevertheless, the daunting stated mission of the society—"to elevate [taxidermy] to a permanent and acknowledged position among the fine arts"—appears to have been accomplished. Even by the third annual exhibition, both the public and scientific communities in newspapers and journals were talking about the new taxidermy as fine art. *Scientific American* noted in 1886 that "the influence of this admirable society may already be seen in the nicer discrimination evinced by museums and collectors in selecting their specimens. It is no longer a question of how much work a man can do in a day in the taxidermist's shop, but rather of the character of his work."[113]

But the society was not merely underscoring contemporary trends, and its founders were not idle supporters of those trends. Rather, the SAT *was* the primary agent of change: its founders created the new American taxidermy movement. In less than five years, the group successfully forced a dramatic paradigm shift, which has not been challenged for more than a century. With the successful public reception of the new taxidermy, natural history museum directors could no longer afford to display synoptic rows of single taxidermied specimens, as museum visitors expected to see more lifelike, artistically prepared taxidermy mounts presented in appropriate habitat settings. Because the society had effected a fundamental change in museum display by accomplishing its main and unstated objective—to "vigorously advocate the group idea"—American natural history museums evolved into educational environments that influenced the way the American public came to view animals and their habitats.

In August 1885, Webster gave an interview to the *Washington Post* about his thriving studio on Pennsylvania Avenue, but was careful to note that "great advances have been made in the art of taxidermy during the past few years, and especially since the establishment of the Society of American Taxidermists, and the art is bound to take a much higher place in the future than has yet been assigned to it."[114] Years later, when Webster's prediction had come to pass, the founders lamented that the society was not better remembered for its contribution to the art of taxidermy—which Hornaday considered to be no less than "breathing new life into stuffed animals."[115] Looking back from the vantage of the late 1930s, Hornaday believed that "the phenomenal rise from that point in the history of museum taxidermy to what it is today has triumphantly vindicated the soundness of the principles on which the SAT was formed."[116] He consoled himself with the knowledge that the practice of simply stuffing animal skins had become "a dead and buried nightmare," and that the growing acceptance of the new methods "had acquired so much momentum that it could not be stopped."[117] Even Lucas, the naysayer of the society, conceded that "men die, institutions pass out of existence, but ideas live."[118]

"The Destruction Wrought by Man": Smithsonian Taxidermy and the Birth of Wildlife Conservation

It is not, perhaps, generally realized how extensive and how rapid are the changes that are taking place in almost the entire fauna of the world through the agency of man. Of course changes have perpetually taken place in the past through the operation of natural causes, and race after race of animals has disappeared from the globe, but there is this wide difference between the methods of nature and man; that the extermination of species by nature is ordinarily slow, and the place of one is taken by another, while the destruction wrought by man is rapid, and the gaps he creates remain unfilled.
—Frederic A. Lucas[1]

With its goal of changing the way museums exhibited specimens, it is perhaps little wonder that the Society of American Taxidermists passed out of existence when it did. Many of the society's most prominent members now had positions of influence within American natural history museums. Of its three presidents, Frederic A. Lucas and William T. Hornaday were now employed at the U.S. National Museum, and Frederic S. Webster owned a private studio in Washington, with the Smithsonian as his primary client. Their unprecedented freedom at the National Museum meant that neither Lucas nor Hornaday had to create experimental mounts at his own expense. However, it also meant that both men were saddled with new responsibilities, and neither contributed any new work to the society's exhibit at the New Orleans Exposition.

Instead, the curators and preparators from every department of the National Museum were expected to contribute to a lavish installation, representing the ethnological, animal, and mineral resources of the entire United States. The undertaking was to be so grandiose that Congress allocated a

Fig. 3.1. William T. Hornaday's sketch for the game animals group to be mounted by the U.S. National Museum for the New Orleans Exposition. In his proposal, Hornaday suggested that game animals "could be mounted on plain pedestals and arranged on the spot in a very striking group (temporarily only) with natural surroundings and effects." (Smithsonian Institution Archives. Image #SIA-2019-006026)

record $75,000 to the exhibit, and a temporary building was constructed to house it until it could be transported to New Orleans. In only seven months' time, Hornaday was expected to prepare specimens representing all the orders of Mammalia in four large cases. Lucas was responsible for preparing companion skeletal mounts—a task so enormous that it eventually proved impossible in the time allotted.[2]

Hornaday proposed an elaborate, multi-tiered display for North American game animals with specimens "mounted on plain pedestals and arranged on the spot in a very striking group (temporarily only) with natural surroundings and effects."[3] His idea, as shown by his pencil drawing, bore remarkable resemblance to Martha Maxwell's terraced group displayed at the 1876 Centennial Exposition in Philadelphia, including a rock outcropping with trees and shrubs placed among male specimens of several species of deer, a moose, a caribou, a mountain goat, a bighorn mountain sheep, a musk ox, an "old" and two yearling pronghorn antelope, and several small mammals. Hornaday doubtless chose to exhibit males of each species because of their impressive size and showy antlers and horns. Maxwell had done the same, but Hornaday departed from Maxwell's design by placing an ambitious family group of bison at center stage.

Frederick W. True, curator of mammals, initially approved Hornaday's plan to include "the entire existing mammalian fauna of North America from the Isthmus of Panama northward,"[4] but Hornaday soon discovered

Fig. 3.2. William T. Hornaday's compromise exhibit of North American game animals for the New Orleans Exposition. (Smithsonian Institution Archives. Image #72-2375)

that the Smithsonian collections did not have mounted specimens of many North American large mammals. His proposed family group of American bison was impossible, as there were no adult males in the collections and only "two, old badly mounted, and dilapidated skins" of a female and a spike bull.[5] New specimens were needed, but Hornaday didn't have the time to seek them out, much less mount them for the exposition. Any improvement of the collection would have to wait until after the close of the exposition. Instead, Hornaday prepared a much modified, less ambitious group, showcasing only the North American ruminants, displayed on a three-tiered, plain white pyramid with none of the habitat accessories he had planned. The public had no way of knowing the compromise this display represented, but they could readily perceive that it did not match the displays of the Society of American Taxidermists in the quality of taxidermy or inventiveness of display.

Over the next several months, Hornaday pushed himself and his assistants to mount and repair specimens for the exposition, focusing his attention on creating groups for the smaller mammalian species, including mink, otters, beaver, rabbits, and squirrels. At the close of the exposition, however, Hornaday and Lucas were asked by George Brown Goode and Secretary Spencer F. Baird to fully identify the deficits of specimens they had encountered in mounting the National Museum exhibit so that appropriate collecting expeditions could be authorized. As it happened, their old colleague from Ward's, Charles H. Townsend, now employed by the U.S. Fish Commission—also under the direction of Baird, who served both as secretary of the Smithsonian and commissioner of fisheries—had recently returned from the West Coast. He carried with him the alarming news that the northern elephant seal now appeared to be extinct.

Hornaday and Lucas agreed that the collecting expeditions should first focus on endangered or recently extinct species while specimens or skeletal remains could still be obtained for research. They also believed that the depredations that endangered such species should be brought to the American public's attention and that compelling exhibits could encourage a more responsible environmental ethic. Goode and Baird concurred. Following Townsend's lead, Hornaday traveled throughout the western United States, particularly Montana, in search of what had fast become the elusive American bison, and Lucas later traveled to Funk Island, Newfoundland, to collect skeletal remains and feathers of the extinct great auk.

Through these expeditions, the U.S. National Museum pioneered the collecting of specimens of endangered animals and skeletal remains of recently extinct species for its scientific collections—and disseminated surplus specimens to museums worldwide to broaden research on these species. But more importantly, these trips served as epiphanies for all three men, guiding the rest of their professional careers—from the scientific and popular articles and books they authored to the societies they founded to the landmark protective legislation they lobbied for. In the end, their efforts would not only transform natural history museums, but would also set the mold for every essential part of the American wildlife conservation movement.

TOWNSEND AND THE ELEPHANT SEAL

On October 14, 1884, Townsend sailed south aboard the schooner *Laura* from the port of San Diego to Magdalena Bay, exploring the coastline and outlying islands in search of the elephant seal. The species was believed to have been driven to extinction by the relentless sealing industry, which had hunted it for its blubber, used to make lamp oil.[6] But then reports began to surface that, from 1880 to 1884, sealers had found and killed nearly three hundred elephant seals along the California coast. A number of the individuals had been sighted at a remote point on the mainland fifty miles south of Cedros Island, marked on all maps at San Cristobal Bay—known among sealers as "Elephant Beach."[7] Townsend had heard of this place from Captain James Morrison of San Francisco, who told Townsend that he himself, aboard the sloop *Liberty*, had visited this beach the previous January and killed thirty-three individuals, and when he returned in March had killed sixty more. Townsend quickly wired the information to Secretary Baird. "The sealers had resumed their destructive work," he later recalled, "and it was a race between us as to whether science or the oil-makers would get the last specimen." Baird wired back that Townsend should charter a schooner and hire Morrison to captain.[8]

When they arrived at Elephant Beach nearly a week later, Townsend found only three juvenile elephant seals sleeping on the sand. He decided not to collect them, hoping instead that their presence would encourage others to haul out at the same spot later in the season. After observing the young seals and recording data for several hours, Townsend selected three members of the crew to stay behind to protect the three from seal hunters and to collect any adults that might emerge on shore in his absence. Townsend continued farther south along the coast for weeks without another sighting. Finally giving up the search, he returned to Elephant Beach, where he found that no adults had turned up, and that two of the three juveniles had left the shore.

He ordered the remaining elephant seal shot and showed the crew how to dissect a scientific specimen that might later be mounted. In preparing this young female, he discovered that the hide "was disfigured by a great gash on the rump, in which the print of shark's teeth were plainly evident." He also found the stomach "terribly infested with abdominal parasites," long, threadlike, white worms, unlike anything he had ever seen in any other pinniped.[9] He collected the parasite for further research.

Realizing how much remained to be learned about the elephant seal, Townsend pressed his search north, again shadowing the coast. He later recalled those months at sea, which had been consumed by hard work and fraught with danger:

> There were the regular watches on deck, which I shared day and night with the small crew; thirsty hunts for wild goats on the mountainous desert islands to replenish our larder; and trips to distant watering-places, where the casks had to be filled and laboriously gotten on board. In our search we must have landed a score of times on rocky islets, inhabited by hundreds of sea-lions, and about which the sea ran high. Day after day we tugged at the oars, minutely examining leagues of beaches while the schooner cruised offshore. We landed through all degrees of surf, where the boat was sometimes swamped.[10]

At one point, the anchor was lost in the rocks; at another, the cast-iron windlass was smashed. Once, a crewman fell overboard in hip boots and nearly drowned. At every turn, the trip seemed to grow more perilous, but Townsend refused to give up.

After weeks of fruitless searching, Townsend ordered the crew back to Elephant Beach. By the time they arrived, it was New Year's Eve—more than two months since their first landing at this spot. Only fifteen elephant seals had hauled out there, including one male, two pups, and twelve females. Townsend could not help but feel discouraged; it appeared that he had arrived a year too late to save the herd. "That a pretty clean sweep had been made of them was evident from the meager results of our own careful search," he wrote at the time. "The great number killed at the old rookery at San Cristobal Bay in the fall and winter preceding our visit was, no doubt, the principal cause of their scarcity."[11]

Townsend spent several hours observing the movements of the elephant seals, taking careful notes, contrasting their methods of crawling with that of black sea lions. Then, when there was nothing left to observe, he instructed the crew to kill all the seals on the beach. Meanwhile, he combed the shoreline, picking through the skeletal remains of other seals. Up and down the coast, Townsend had found weatherworn skulls and bones, evidence, he wrote, "that their former abundance has not been overestimated."[12] That day, on Elephant Beach, he found a tide-beaten skull that measured almost exactly two feet across, which Townsend calculated would have come from a male nearly twenty feet long—twice the size of any they had observed. Surveying the scene, he concluded that "this interesting and valuable animal has heavy odds to encounter in its struggle for existence."[13]

HORNADAY AND THE LAST BISON HUNT

Keen to raise awareness about extinction and endangered species through public exhibition, Hornaday became convinced, like Townsend, that he was in a race against the hide hunters. While preparing the North American mammal exhibit for New Orleans, he discovered that the Smithsonian had only two poorly mounted bison specimens. Hornaday made plans to collect additional specimens and requested permission from Secretary Baird to organize a trip to Montana to collect bison for research and to raise public awareness about the animal's demise through the exhibition of these "most valuable and interesting American mammals."[14] Baird agreed, and instructed Hornaday to collect twenty to thirty complete bison skins and skeletons, and at least a further fifty skulls. These specimens would complete the National Museum's own collection and also allow the Smithsonian to supply specimens to museums worldwide.

Historians have long argued that the decimation of the species was primarily the result of a racist U.S. government policy that encouraged hide hunters and the military, engaged in battles with Plains people, to remove their main food source so as to speed their resettlement on reservations. However, there appears to be no historical evidence to support this claim. Dan Flores writes in *American Serengeti* that the bison hunter James H. Cook, in *The Border and the Buffalo*, his account of the hunt published in 1907, attempted to glorify the hide hunters by claiming that they were a part of a government-sanctioned policy, devised by General Philip H. Sheridan, to rid the prairie of Plains people. Sheridan would have patterned such a policy after his successful Civil War campaign in the Shenandoah Valley, in which he commanded Union troops to lay waste to the valley, burn crops and barns, and seize livestock. Although Sheridan's strategy ultimately destroyed the South's food supply and hastened the end of the war, there is no evidence the same strategy was deployed as an explicit policy against Plains tribes. It appears that twenty years after Hornaday published *Extermination of the American Bison,* in which he blames hide hunters for pushing the species to the brink of extinction, Cook fabricated the Sheridan policy to exonerate the hide hunters from this horrible legacy.[15]

Only a hundred years earlier, tens of millions of bison had roamed the Great Plains of North America. Lewis and Clark described the herd as a "moving multitude" that "darkened the whole plains." But in the final decades of the 1800s, cultural change swept across the prairie, devastating Plains people and pushing bison to the brink of extinction. That change arrived in the form of drought, Euro-American settlers, horses, railroads, diseases, and commercial hunting. After the Civil War, the railroads brought west thousands of former Civil War soldiers turned hide hunters and facilitated the transport of bison skins, bones, and tongues (a delicacy) to coastal markets. The need for strong belts to turn the millworks fueling the Industrial Revolution ensured endless demand for leather.[16] Hornaday would have to act quickly to secure research and taxidermy specimens for the Smithsonian.

In early spring 1886, arrangements were made for Hornaday to travel to Miles City, Montana—the center of the range of the vanishing northern herd.[17] This was not the best season to hunt bison because they would be molting their winter coats, but Hornaday decided to make an initial trip to discover the whereabouts of any remaining individuals, if there were any to find. The secretary of war ordered the officers of Forts Keogh, Maginnis, and McKinney in the Montana and Dakota Territories to furnish Hornaday with a collecting party equipped with supplies, and the secretary of the interior ordered Indian agents and scouts to assist him when called upon to do so. Hornaday, his assistant Andrew Forney, and George Hedley, an acquaintance and taxidermist from New York, arrived in Miles City in May. Captain J. C. Merrill, a doctor in the U.S. Army stationed at Huntley—who regularly collected specimens for the National Museum—had written to Hornaday, informing him that local rumors suggested bison could be found near Big Dry Creek. So the Smithsonian Institution Buffalo Outfit traveled northward along the Missouri River, and it was there that they witnessed "where the millions had gone":

> The bleaching skeletons lay scattered thickly all along the trail. Like ghastly monuments of slaughter, . . . they lay precisely as they fell four years before, except that the flesh was no longer upon them. The head stretched far forward as if for its last gasp. . . . The skinners always left the heads of the bulls unskinned, and the thick hide had dried down upon the skulls harder than the bone itself. . . . Many of these heads were

so perfectly preserved, and with their thick masses of wavy brown hair were so fresh looking, that the slaughter of the millions was brought right down to the present, and seemed to have been the work of yesterday. We could endure the sight of the bones reasonably well, . . . but these great hairy heads made us feel our loss most keenly.[18]

Failing to find any signs of live bison, the party moved southwest toward the LU-Bar Ranch, where the owner had reported seeing a herd of thirty-five. Once Hornaday had set up permanent camp, Irvin Boyd, a Montana cowboy who worked at the ranch, and a Cheyenne Indian scout joined the party to serve as guides. Two days later, they came upon a solitary bison calf. After another week, they found two bull bison on Little Dry Creek, but only managed to capture and kill one of them. This animal was still shedding its winter coat, and because the skin would not make a representative mount, they took only its head and skeleton. Hornaday cut the trip short, as the calf and the few adult bison they had seen proved conclusively that there were indeed a small number of wild bison remaining and breeding in Montana Territory. Forney went back to Washington with the bison calf, and Hornaday hastily packed up the numerous skeletal specimens they had collected and followed the two back to the Smithsonian.[19]

In July 1886, large crowds gathered on the lawn of the National Museum to see the bison calf. Hornaday had nicknamed him "Sandy" owing to his "luxuriant growth of rather long, wavy hair, of a uniform brownish-yellow."[20] But this crowd-pleasing calf was still too young and weak to roam freely over the lawn in front of the museum; he was kept tied to a stake during the day and brought into the taxidermy workshop at night to prevent him from being stolen.[21] Many feared that the sickly calf would not survive. In fact, Sandy was so docile that Hornaday even posed for a photograph with him, loosely holding the picket rope knotted around the animal's neck. Within a few weeks, however, Sandy began to improve, and soon he became too strong and unruly even for Andrew Forney, who was in charge of caring for him.

One night after a hard rain, as Forney led Sandy toward the museum's taxidermy workshop, the calf began running down the slope, "head down and tail in the air, the mud flying from his heels," Hornaday wrote. "After him raced Andrew, hanging helplessly to the rope."[22] Clearly, there was not enough space for Sandy at the museum, so Hornaday took him to the country residence of Newton P. Scudder, the museum's librarian, where he was turned out to pasture to graze with cattle. Within a few days, Sandy was

Fig. 3.3. William T. Hornaday with Sandy, 1886. (Smithsonian Institution Archives.
Image #79-13252)

dead. At first, shocked by the calf's sudden death, Hornaday suspected that
the animal had been poisoned, but it soon turned out that Sandy had died of
pasture bloat after eating damp clover.[23] The *Washington Post* reported that
Hornaday "was for a time almost inconsolable at its loss."[24]

Hornaday's devastation was more than fatherly affection: by now, he
was keenly aware that the American bison was headed toward extinction.
Every death brought that fate a step closer. He skinned Sandy and, using the
photograph as a guide, mounted him in a lifelike pose. Until the specimen
joined the Smithsonian bison group in the museum, Hornaday placed it at

the entrance to the taxidermy workshop—an emblem of the rapid extinction continuing far from the nation's capital and a daily reminder of the urgency of their task.[25] As Hornaday and his assistants prepared for a return trip to Montana, they were determined to preserve some vestige of the American bison—if only as mounted specimens.

Hornaday and the Buffalo Outfit returned to Miles City in September. It took three months of hunting to find and kill twenty-two bison, and to collect two dozen skins, sixteen skeletons, and fifty-one skulls from dry remains.[26] So depleted were the once great herds that when Hornaday came upon a group of fifteen bison, he decided to leave them. The irony of killing the "last" of a species was not lost on him. Upon his return to Washington, Hornaday hatched an elaborate plan to save the last of the wild bison herd.

Fig. 3.4. Andrew Forney (*seated left*, preparing a Bengal tiger skin), William T. Hornaday (*at center*, mounting Bengal tiger), and an unidentified assistant are shown working in the taxidermy workshop at the U.S. National Museum, circa 1886. Note the bison skull and skin in the foreground. (Smithsonian Institution Archives. Image #MNH-2783)

LUCAS AND THE GREAT AUK EXPEDITION

The possibility that many of North America's disappearing species might be preserved only as taxidermied specimens was growing as grimly apparent to Frederic Lucas as it was to Hornaday. As early as 1885, Lucas discussed with Secretary Baird the possibility of sending a collecting expedition to Funk Island, off the northeastern coast of Newfoundland, once the breeding ground for the great auk (today known as the "Great Auk Graveyard"). As osteologist, Lucas discovered that the museum possessed only one poorly taxidermied specimen, an egg, and a single humerus of the once plentiful bird. The great auk was long extinct, and Lucas held out no hope of finding any living vestiges of the species; however, he felt certain that numerous bones could be collected on Funk Island and that a skeleton could be constructed from those remains. Unfortunately, at the time, the museum was not prepared to invest the considerable amount of time and money that such an expedition would require—especially while it was focusing on species that had not yet gone extinct.[27]

For the next two years, while Hornaday turned his efforts toward obtaining specimens of North American mammals, particularly the American bison, Lucas concentrated his attention on assessing the museum's collection of endangered species. He undertook a report on the world's species that were either extinct or threatened with extinction, warning readers of the rapid changes occurring in the world's fauna by the "agency of man." Lucas enumerated the "more obvious causes of extermination," including agriculture, the increase of domestic livestock and the perceived need to protect herds from predation, introduction of non-native species, and over-hunting of economically valuable species for food, fashion, or sport. These causes, Lucas believed, were a direct result of "the common fatal fallacy that because some animals exist in large numbers, the supply is unlimited and the species needs no protection, a belief that is usually acted upon until the species is verging on extinction."[28] The growing importance of this work was punctuated by Hornaday's experience in the field, which raised concern not only among the scientific community, but also the informed public, many of whom did not agree with the National Museum's collecting of endangered species.

In March 1887, Lucas found himself defending the museum, and the work of Townsend, when he wrote a response to an editorial titled "Official Extermination," published in *Forest and Stream*. In the years preceding, the northern elephant seal had been extirpated from the California coast and nearly hunted to extinction. The author of the editorial rebuked the National Museum and Townsend for procuring specimens, calling such hunting "cold-blooded, remorseless and heartless."[29] This editorial underscored the growing public concern for the conservation of wildlife, but it also suggested a lack of understanding about the role that natural history museums were playing in the preservation of threatened species. Lucas wrote in Townsend's defense, commending him on "having secured for science even a few immature individuals of this, our largest pinniped." It was better, he argued, to obtain "all the specimens possible for scientific purposes, although at risk of exterminating the race," than to leave these survivors "to the tender mercies of the seal hunters."[30]

Furthermore, Lucas explained that the National Museum, after selecting a few individuals for its own collection, had distributed the remaining specimens to the British Museum, the Museum of Comparative Zoology, the American Museum of Natural History, and the Academy of Natural Sciences of Philadelphia. With the total loss of the great auk, both species and specimens, in the forefront of his mind, Lucas asked whether the author of the editorial "would prefer that these skins should have been made into leather and their bones left to whiten on the shore. . . . No one deplores the destruction of animals more than does the present writer, and yet he deems the slaughter of the sea elephants not only justifiable but commendable." Appropriately, Townsend's own further explorations pushed the northern elephant seal back from the brink of extinction: while on expedition in 1911, he again discovered a surviving colony of the seals on Guadalupe Island, which led the Mexican government, and later the U.S. Congress, to pass protective legislation.[31]

The great auk would experience no such resurrection. Nevertheless, Lucas jumped at the opportunity to join the U.S. Fish Commission's new schooner *Grampus* on an expedition to the coasts of Newfoundland and Labrador in summer 1887. Along with his observations of the history and comparative anatomy of the great auk, Lucas's report of the expedition included documentation of individual, secondary, and age variation in this population. But it also reads like much of the eloquent prose of the famed American naturalists John Burroughs and John Muir. Of his first impression of Funk Island, he wrote: "A large portion of the southern and most extensive swell of rock is thickly covered with vegetation, this, the former breeding ground of the great auk, being mapped out in vivid green by plants nourished by the decomposed bodies and slowly decomposing bones of the long extinct bird."[32] Lucas read the abundance of vegetation to locate the thick tangle of millions of auk bones that lay just below the sod. Near the rock "pounds" or pens, where the birds were "driven like so many sheep"[33] and kept until they could be slaughtered, were found the most abundant remains—Lucas imagined the scene where "kettles once swung in which the birds were parboiled to render plucking them an easy operation."[34] The collecting party removed sections of the sod that were ten to twelve feet in diameter and two inches deep to reveal a compact layer of charcoal and bones. Not a skull was found that did not have a break across the top or the back "entirely lacking," as the birds were clubbed about the head by the "feather-hunters."[35]

Although the expedition succeeded in collecting thousands of bones of the extinct bird, Lucas noted that in the end they "made up not more than a dozen skeletons, and these not absolutely perfect."[36] However, he later indicated that he was able to piece together at least five "perfect specimens," meaning that they were constructed from the bones of various individuals: one remained in the National Museum, others were sent to the Museum of Comparative Zoology at Harvard, the AMNH, the Museum of Science and Art in Edinburgh, and the Australian Museum in Sydney. The National Museum also kept two skeletons for its "reserve series," as well as a variety of individual bones for scientific study. In 1890, Lucas's anatomical findings led to the remounting of the one great auk already in the museum's collection, shortening it by three inches, as most early mounting techniques tended to stretch the skin.

The *Grampus* expedition awakened in Lucas a profound understanding of the dangerous pace at which species were becoming extinct. He remembered observing on voyages with his father to Asia, Africa, and South America that there appeared to be an endless abundance of various species, particularly birds, but in a few short decades he bore witness to dramatic changes in their numbers. The great auk was a harbinger of the fate of the countless species threatened by civilization.

Determined to bring this problem to the attention of the scientific community and the American public, Lucas began publishing numerous scientific and popular articles in publications such as *Popular Science Monthly*, *Nature*, and *The Auk*. He also completed his assessment of the Smithsonian's holdings of extinct and endangered species in both the scientific and exhibition collections. In the Smithsonian's annual report for 1889, he published an annotated list of fourteen specimens that represented species recently extinct or threatened with extinction, including the great auk, dodo, Labrador duck, West Indian seal, "California sea elephant" (northern elephant seal), Pacific walrus, Steller's sea cow, Galapagos tortoise, and tilefish. It was the first work of its kind—but it was not merely a dispassionate enumeration. Lucas, always the voice of reason, called upon naturalists and the general public to work together to preserve species from destruction.[37] He would find no stronger ally than William T. Hornaday.

THE DEPARTMENT OF LIVING
ANIMALS AND THE BISON GROUP

In the early months of 1887, Hornaday learned that hunters had killed three members of the last band of bison he had left behind in Montana.[38] Something needed to be done. Hornaday proposed two parallel objectives to Goode.

First, recalling Sandy's popularity with the public, he suggested that the Smithsonian create a zoo—or, as he called it, a Department of Living Animals—with the intention of establishing a captive breeding program for the last remaining bison. Goode approved the idea early in 1887, and by October the new department was formally organized, with Hornaday as its curator. Goode justified the move by claiming that the department would "afford to the taxidermists an opportunity of observing the habits and positions of the various species, with a view to using the knowledge thus acquired in the mounting of skins for the exhibition series of mammals."[39]

Second, Hornaday proposed an ambitious bison group that would represent the animals in their native Montana habitat, with soil and vegetation that he had collected out West—a dramatic change from earlier, more static natural history displays. He hoped with this exhibit to introduce the species to museum visitors and, in turn, raise their awareness of the plight of his beloved bison. But also, should the Department of Living Animals fail, this group might prove to be the sole lasting record of this once great species.

The Department of Living Animals, which started with two bison fenced behind the Smithsonian Castle, evolved into the National Zoo, and the National Museum became the first American museum to exhibit a family group of taxidermied bison. The exhibit represented the very height of scientific taxidermy.

With these dual goals in mind, and a mounting sense of urgency, Hornaday organized an expedition to the Northwest to collect numerous species for the Department of Living Animals. Secretary Baird made arrangements for him to accompany the U.S. Fish Commission's special freight car. After its cargo of live fish were distributed on the initial trip west, the car was modified to hold live animals upon its return. On October 8, 1887, Fish Commission Car No.1 left Washington, D.C., for the Pacific coast— its scheduled stops included St. Paul, Minnesota; Fargo and Mandan, Dakota Territory; Helena, Montana Territory; Tacoma, Washington Territory; Portland, Oregon; Mountain Home, Idaho Territory; Salt Lake City, Utah

Territory, and Cheyenne, Wyoming Territory. As word spread that the National Museum was collecting live animals, citizens flocked to the train stations either to offer specimens as gifts or to sell them. The species that were to form the nucleus of the Department of Living Animals included two red foxes (*Vulpes vulpes*), a cross fox (a partially melanistic color phase of the red fox), a cinnamon bear (*Ursus americanus cinnamomum*), a white-tailed deer (*Odocoileus virginianus*), a Columbian black-tailed deer (*Odocoileus hemionus columbianus*), a mule deer (*Odocoileus hemionus*), two American badgers (*Taxidea taxus*), a spotted lynx or bobcat (*Lynx rufus texensis*), four prairie dogs (*Cynomys ludovicianus*), and a golden eagle (*Aquila chrysaetos*). Having traveled over seven thousand miles, the freight car returned to Washington, D.C., on November 8 with a sizable collection. In short order, the animals were housed in cages in a 25-by-106-foot heated wooden enclosure—built from materials salvaged from the New Orleans exhibition annex—on the south side of the National Museum building.[40]

Hornaday wasted no time in drawing attention to his small menagerie. "The whole movement has been prompted by the fearful rapidity with which game is being killed in the West and in the absolute certainty that in a few years many of the representative American animals will be entirely extinct," he told the *Washington Star*.[41] In the end, Hornaday's efforts paid off, as the newspapers encouraged attendance.

On December 31, 1887, the small zoo was opened to the public. By the end of January, the number of animals had increased to a total of fifty-eight mammals and birds. On February 1, Hornaday hired away another Ward's employee, Nelson R. Wood, a highly skilled bird taxidermist and artist, as the collection's keeper. Wood later served in the National Museum's taxidermy division until his death in 1921. In spring 1888, six bison were donated to the museum, including a cow and a bull bison—purchased for the Smithsonian by New York Fish Commissioner Eugene G. Blackford from a rancher in the Sandhills of western Nebraska, where they had been reared with cattle and were quite docile. The bison became the department's most popular attraction. Energized by this swift achievement, Hornaday proposed to Goode in early December 1888 that the Smithsonian assume the responsibility of establishing a captive breeding program for the bison.

But, for the present, Hornaday turned his attention toward designing and mounting his group of bison, which he viewed as a taxidermal "experiment [that] was to be regarded as a crucial test of the group idea as adapted to the purposes of scientific museums."[42] By now, mounted groups were proliferating in American natural history museums. If they were not yet

Fig. 3.5. Cow and bull bison from Nebraska housed in the temporary enclosure on the south side of the U.S. National Museum, circa 1888. (Smithsonian Institution Archives. Image #MNH-8008A)

dominant, they were certainly not uncommon. So what about Hornaday's bison group did he expect to be so revolutionary? The answer lies in the distinction he made between what he termed a "special exhibition group" and a scientific group. Jules Verreaux's "Arab Courier Attacked by Lions," Edwin Ward's "Lion and Tiger Struggle," and John Wallace's "Lions Fighting" all epitomized the "special exhibition group": all depicted species locked in life-and-death struggles, dramatized with bloody wounds.[43] In short, Hornaday's idea of what was appropriate for a natural history museum was evolving. Perhaps now he agreed with Webster's earlier assessment that "A Fight in the Tree-Tops" was too sensational for the scientific museum. Because exhibition groups were "theatrical in effect," he now felt they were best suited for "great expositions, for show-windows, fairs, crystal palaces."[44]

He codified this division into a set of rules, beginning with an admonition to taxidermists to "suppress all tendency to the development of violent action on the part of your specimens":

In a well-regulated museum no fighting is allowed. Represent every-day, peaceful, home scenes in the lives of your animals. Seek not to startle and appall the beholder, but rather to interest and instruct him. . . . Let them be feeding, walking, climbing up, lying down, standing on alert, playing with each other, or sleepily ruminating—in fact, anything but fighting, leaping, and running.[45]

While his own second group, "The Orang Utan at Home," followed these new rules strictly, it raised for Hornaday another troubling issue: it was in fact a scientifically inaccurate grouping. Hornaday had designed the exhibit to show the range of secondary sexual variation and age variation in the species, but in so doing he sacrificed the accurate portrayal of orangutan behavior. He well understood that they are a solitary species, as he described in a scientific paper he delivered before a meeting of the AAAS and again in his popular book *Two Years in the Jungle,* and thus would never be found in a large group in the wild. In an attempt to deemphasize dramatic violent scenes, Hornaday and other taxidermists promoting the group idea had instead accentuated "domestic cheerfulness."[46] Now, at last, with the mounting of his bison group, he could design a large mammal group that would be scientifically accurate—pushing him to demand increased accuracy in all parts of the exhibit.

No deception would be tolerated to augment or in any way manipulate the appearance of the specimens. It was then common practice to exaggerate the size of any large male specimen by stretching the hide or overstuffing, but Hornaday renounced such trickery, insisting that "we endeavored by every means in our power, foremost of which were three different sets of measurements, taken from the dead animal, one set to check another, to reproduce him when mounted in exactly the same form he possessed in life."[47] Lewis L. Dyche, professor of natural history at the University of Kansas, was apprenticing under Hornaday, learning to mount the bison so that he might return to the university's newly erected natural history museum to prepare his own group. When Dyche later sent photographs of the bull he had mounted in Kansas, Hornaday was dismayed. Dyche had departed from anatomical accuracy, making the bull thinner to accentuate its musculature.

"While I recognize the fact that this was quite intentional," Hornaday wrote to Dyche, "I think it hardly does the animal justice. The typical buffalo must be a well-fed animal, though not necessarily *fat* by any means." Worse still, Dyche had mispositioned the bull's pelvis. "I think I see exactly how it happened," Hornaday wrote. "You got so interested in the development of the muscular anatomy you forgot the osteological side of the problem." To correct this now would mean extensive reworking of the mount, but Hornaday considered the error too egregious to leave uncorrected. He advised Dyche to "cut it open and change that even now!"

Hornaday's newfound rigor extended to all parts of his bison exhibit. The bottom of the case was covered with pieces of "genuine prairie sod, each about one inch thick and a foot square, cut on the buffalo range in Montana, and shipped in barrels to Washington."[48] Inside the case, the sod squares were "matched carefully" and "the joints . . . skillfully closed."[49] Likewise, the exhibit featured "clumps of sage brush and bunches of broom sedge, grubbed up in Montana," a bison trail created with "Montana dirt," tracks from "genuine buffalo hoofs," and some bison bones as "often seen protruding from the faces of cut banks in Montana."[50] Hornaday bragged that of "all the accessories in the buffalo case, everything in sight came from the Montana buffalo range, except the sheet of glass forming the surface of the pool."[51] Before its unveiling at the opening of the National Museum's Hall of Mammals, the *Washington Star* echoed his excitement, announcing that the exhibit would feature "real buffalo-grass, real Montana dirt, and real buffaloes."[52]

While Hornaday had long advocated scientific accuracy in taxidermy, the bison group was a new pinnacle. With an eye toward the possibility that bison, as a species, might soon become extinct, Hornaday understood that these specimens needed to be exactly accurate. In fact, he designed the arrangement of the group with this in mind—placing the adult bull and cow in the foreground, where they could be easily viewed, and positioning them on level ground. Hornaday explained:

> If the huge bull bison in our large group had been put walking up hill, or walking down hill . . . his height at the shoulders would be either exaggerated or diminished, almost unavoidably. As it is, he was with deliberate intention mounted on a flat and horizontal surface, as was the cow also, so that even though they are in a group they lose nothing whatever of their value to the technical zoologist, who demands that all specimens shall be mounted on flat surfaces, and in conventional attitudes for the sake of comparison.[53]

Yet Hornaday's confidence, or perhaps arrogance, appears to have annoyed his colleagues and superiors at the museum. One year before the bison group was unveiled, the Washington correspondent of the *Chicago News* chided Hornaday for touting his bison bull as "a great work of art, as well as a true reproduction of nature" and for his excessive pride in "his

Fig. 3.6. The U.S. National Museum bison group mounted by William T. Hornaday
between 1886 and 1887. (Smithsonian Institution Archives. Image #NHB-5470)

success as well as his skill."[54] Convinced that the experts on living bison
would agree, the correspondent wrote, Hornaday placed the bull on a pedestal at the museum's entrance and invited Spencer Baird, General Sheridan,
and General Van Vliet, along with "a number of other distinguished army
officers, who had chased the bounding bison over the plains," to critique
his work, "expecting to hear nothing but eulogisms." Hornaday was stunned
when

> to his disappointment and dismay they all, with one accord, commenced
> making the most savage criticisms. One said it was too short, another
> thought it was too long; others claimed it was too fat, more that it was
> too lean. Some said the position of the legs was not natural, and several
> declared that no living buffalo ever stuck his nose up in the air like that.
> There was not a hair that pleased any one. The entire company expressed
> their surprise that a man of Mr. Hornaday's experience and skill should
> waste his time stuffing rusty old bison like that one. He was advised to
> throw it away and take another trip to the West to get a good one.[55]

"Poor Hornaday," the correspondent wrote, "was all broken up." But in truth, he explained, Hornaday was the victim of a practical joke, a conspiracy by the invited critics "to humiliate him." One of the Smithsonian's scientists had confided, "Hornaday was getting altogether too much glory out of his condemned old beast, and we thought we would fetch him down a peg. He had actually convinced himself that there was not a specimen in the whole museum worth looking at except that bison. He thinks differently now."

Hornaday, thin-skinned and prideful, refused to see the humor. He was outraged by the publication of the article—which was widely circulated. He wrote to the *Boston Advertiser*, where it was reprinted, to decry it as "a series of malicious falsehoods" and the Smithsonian's "scientist" [Hornaday's quotation marks] as "an infernal liar." In his own defense, Hornaday supplied a copy of the letter submitted by General Van Vliet to Baird rendering his true opinion. Van Vliet wrote:

> Gen. Sheridan thought the animal was too tall but the taxidermist showed us, in his notebook, the measurements he made of the animal when he shot him, and they agreed with the stuffed animal. I thought that the left hind leg might be brought forward six inches. This would make the animal look a little shorter; but I doubt if I would even do this. It is a magnificent specimen as it is, and perfectly natural.[56]

The editors of the *Advertiser* were swift to apologize, explaining that they never understood the quality of Hornaday's work to be legitimately in question, and that this was "a practical joke upon Mr. Hornaday" that "was done very gravely, as the report stated, naturally perplexing and vexing the artist." But, having inadvertently offended him, the *Advertiser* officially assured Hornaday that they were certain his bison deserved Van Vliet's praise "if the mounted bison is equal to the other specimens of his work which adorn the national museum."[57]

Indeed, Van Vliet's praise was the general consensus—which was only magnified by the completion of the entire group. George Browne Goode praised Hornaday's work for its artistic effect and scientific accuracy, as "a triumph of the taxidermist's art, and, so far as known, it surpasses in scientific accuracy, and artistic design and treatment, anything of the kind yet produced."[58] R. W. Shufeldt, in his critique of the National Museum's taxidermy, praised Hornaday's bison group as "one of the very finest accomplishments that the art of taxidermy has produced in this country."[59] To emphasize the group's artistic merits, he compared it to Paulus Potter's famous painting of a young bull, saying, "And were I to choose between

Fig. 3.7. William T. Hornaday's "Mammal Extermination Series" exhibit displayed at
the Centennial Exposition of the Ohio Valley in 1888. (Smithsonian Institution Archives.
Image #MNH-4465)

being the author of Paul Potter's bull and these buffalo, I should without a
moment's hesitation decide in favor of the latter."[60]

After completing the bison group, Hornaday undertook a special exhibit
for the Centennial Exposition of the Ohio Valley, called the "Mammal Extermination Series." Unveiled in 1888, the exhibit included the American
bison, moose, elk, antelope, mountain goat, mountain sheep, walrus, elephant seal, and beaver. Its focus, however, was the American bison. In the
center of the display was a stark case:

> On a section of Montana prairie, eight feet by ten, lies the complete
> skeleton of a large buffalo bull, just as it was found bleaching on the
> range, and just as ten thousand others lie to-day. The powerful action of
> the weather has stripped every particle of flesh from the bones, and left
> them clean and white, but still, attached to each other by their dried up
> ligaments, the legs in precision precisely as the animal fell.[61]

Forest and Stream called it a "ghastly object" that "surely must awaken
a feeling of remorse in the breast of every old buffalo hunter." The exhibit
included hides of various market values, instructive maps, oil paintings by

James Henry Moser, and weapons used in the destruction of currently endangered species. Its effect was "both impressive and saddening to every
lover of animated nature."[62] Indeed, the "startling" exhibit stood in stark
contrast to others of the period. Sociologist of science Susan Leigh Star, in
her critique of taxidermy, makes the overarching claim that "taxidermy
has cleaned up the mess of colonialism, patriarchy, and violence against nature."[63] Yet Hornaday's exhibit was a graphic demonstration of the effects of
colonialism on wildlife, as it highlighted the violent deaths of millions of
American bison at the hands of white hunters.[64]

In 1889, Hornaday took his message of the impending extinction of bison to a larger audience with the publication of *The Extermination of the
American Bison*, considered by historians to be the first important text of
the American wildlife conservation movement. While John Muir, the famed
American naturalist, was arguing for the preservation of natural habitats—
such as Yosemite National Park, established in 1890—Hornaday recognized
that species such as the American bison faced more imminent extinction.

BIRTH OF A NATIONAL ZOO—AND
THE OUSTER OF HORNADAY

Hornaday's expeditions to Montana made him determined to raise public
awareness of the need to establish legislative protections for threatened and
endangered species and, in the most extreme cases, to undertake captive
breeding programs to redress decades of reckless slaughter. Through his per-
sistence, the Smithsonian's Department of Living Animals was eventually
reconceived as the U.S. National Zoological Park, and his lobbying efforts
persuaded the Fiftieth Congress of the United States (1889) to allot $200,000
to establish the zoo, with the stated purpose of effecting "the preservation
and breeding in comfortable, and so far as space is concerned, luxurious
captivity of a number of fine specimens of every species of American quad-
ruped now threatened with extermination."[65]

On May 10, 1890, Hornaday was appointed acting superintendent of the
National Zoological Park.[66] However, the new secretary of the Smithsonian,
Samuel P. Langley, disliked Hornaday's newfound influence and worked to
ensure that the superintendent of the zoo would be little more than a care-
taker with no freedom to guide the direction of the new park. After much
rancor between the two men, Hornaday resigned his positions at the zoo
and museum in June. It was a critical turning point in Hornaday's professional life, marking the end of his long career in museum taxidermy and
the beginning of his passionate struggle to protect America's wildlife. Many

years later, Hornaday wrote that "it was one of the wisest acts of my life . . . speed[ing] my progress toward the theatre of my real life work."⁶⁷

Despite Hornaday's premature departure from the National Zoo, his dream of breeding bison in captivity was soon realized. On November 3, 1890, the female bison from Nebraska that had been donated to Hornaday's experimental zoo gave birth on the lawn of the National Museum. Hundreds of visitors flocked to the zoo each day to see the infant bison. This yellow-brown newborn looked much like Sandy, the calf with whom Washingtonians had fallen in love only four years earlier.

For the next five years, Hornaday lived in Buffalo, New York, working in real estate and dabbling in local politics, until the Depression of 1893—one of the worst in American history—hit the country hard, throwing the real estate business into a slump. At the depth of the financial crisis, Hornaday received an unexpected letter from Henry Fairfield Osborn, then professor of zoology at Columbia University, which had been prompted by a strong recommendation from C. Hart Merriam, chief of the Division of Biological Survey, U.S. Department of Agriculture, asking Hornaday to direct a "New York zoological park—worthy of the metropolis of the Western Hemisphere," which had recently come under the auspices of the newly formed New York Zoological Society (NYZS; now the Wildlife Conservation Society).⁶⁸

After his experience at the Smithsonian, Hornaday was wary of "'scientists,' and zoological parks." He told Osborn, "If you wish for director a zoological investigator, then I am not your man for the place. If you wish a practical administrator who knows a lot of practical zoology, I think I could give you good help." Given that the society's board members were predominantly lawyers and businessmen who shared a sportsman's desire to preserve wild game while both entertaining and educating the public (all were members of Theodore Roosevelt's Boone and Crockett Club), it is not surprising they insisted that Hornaday was just the right man for the job. For his part, he saw this position as an opportunity to resume his crusade for wildlife preservation and pursue his dream of breeding endangered species in captivity at an institution that afforded him unmatched resources and freedom. The New York Zoological Park, soon popularly known as the Bronx Zoo, was ready for him to design from the ground up. On April 1, 1896, Hornaday officially accepted the position as its first director.⁶⁹

In an interview published in the *New York Times*, Hornaday addressed his New York audience. In a flourish of optimism, he declared that the zoological park "will become the most popular recreation of Greater New-York." He also proposed a pedagogical component devoted to "popular education in zoology" that would emphasize the need for species preservation, and instruction in the fine arts focused on the sketching, painting, and sculpting of animals, where artists and especially taxidermists would be allowed to study living animals—just as he had intended for the National Zoo. Hornaday also seized this opportunity to transform his public image from "hunter, naturalist, taxidermist" to administrator and spokesman for wildlife preservation. He asserted that he had "never been what you might call a sportsman, for while I have killed scores of species and hundreds of individuals of large game animals, I have never hunted save as a naturalist, bent on making studies and preserving in one form or another every animal killed that was worthy of a place in a museum." He went on to explain that while he enjoyed hunting, and was "still savage enough to enjoy stalking a fine, keen-witted animal," he was violently opposed to the so-called sportsmen who hunted for the sake of bagging a large quantity of game. Hornaday made it clear that he intended to use his new position to vigorously promote wildlife preservation.[70]

Two months later, Lucas and Townsend were appointed to the second Joint High Commission of the Fur Seal, composed of scientists from the United States—led by influential ichthyologist and Stanford president David Starr Jordan—and Great Britain, to investigate the existing condition of the fur seal herds in the Bering Sea. At the time of his appointment, Jordan was pleased to discover that "two of the ablest naturalists of the United States National Museum," Frederic Lucas, now curator of the Department of Comparative Anatomy, and Leonhard Stejneger, curator of the Department of Reptiles, were appointed as associates, along with Charles Townsend, naturalist of the U.S. Fish Commission steamer *Albatross*, which had been placed in service to the fur seal commission. While Lucas focused his research on observations aimed at solving certain disputed points of anatomy, Townsend mapped and photographed the rookeries, and together they cruised among the sealing fleet and dissected seal carcasses on the deck of the *Albatross*.[71] The significance of the appointment for both Lucas and Townsend was far-reaching, as it secured for each status in the scientific community as "naturalists of highest standing," in the words of David Starr Jordan.[72] This status, coupled with their experience in museology, quickly propelled them into influential administrative positions alongside their friend and colleague William T. Hornaday. Though they could not know it at the time, the research they would gather about the northern fur seal and the recommendations they would make for managing the herd would lead them into direct conflict with Hornaday and his renewed campaign for wildlife preservation.

CHAPTER FOUR

Competing Ideas, Competing Institutions: Decorative versus Scientific Taxidermy at the Carnegie and Field Museums

Taxidermy, to-day, as I see it, is divided into two distinct classes, viz., Museum or Institutional Taxidermy, which is conservative and settled, and Decorative Taxidermy, which is radical and progressive. To one or the other every taxidermist . . . must belong.
—Frederic S. Webster[1]

During the 1890s, major metropolitan areas across the United States experienced a boom in the establishment of new museums, but no two were more important or representative than the Field Museum of Natural History (FMNH) in Chicago and the Carnegie Museum of Natural History (CMNH) in Pittsburgh.[2] Each was founded with private money from a single philanthropist—Marshall Field in Chicago, Andrew Carnegie in Pittsburgh—who wished to build for his city a crown jewel of culture and enlightenment, an institution representing both the arts and the sciences. Designed for public edification and enjoyment, both of these museums were monumental structures with capacious interiors dedicated to innovative public exhibitions.

Not surprisingly, to fill these grand new exhibit halls, the directors of both museums sought to hire the finest taxidermists. The Field Museum hired Carl E. Akeley and the Carnegie hired Frederic S. Webster—both former foremen of the taxidermy department at Ward's Natural Science Establishment with additional years of experience mounting exhibits for museums. Despite their common training, however, the two men left Ward's with competing ideas about taxidermy. Akeley believed that successful museum taxidermy should seamlessly join art and science, while Webster believed that taxidermy was a decorative art and that the museum's traditional emphasis on scientific collections served only to confine the taxidermist's artistic expression. As a result, though both men were remarkably skilled practitioners, only one had the vision to bring museum taxidermy into the twentieth century.

DECORATIVE AND SCIENTIFIC TAXIDERMY

After Frederic A. Lucas left Ward's for the U.S. National Museum in 1881, Henry A. Ward had difficulty finding the right person to replace him as foreman of the establishment. In a letter to Ward, William T. Hornaday offered his opinion and advice:

> I think Webster would be a highly capable man. . . . For he has an artist's eye for form and proportions, and he knows *all* the methods and tricks of the business. . . . As to the scientific work—particularly the determination of species—Webster will have to *learn* it, but he will take it *quite* as readily as anyone with you.[3]

Hornaday feared it would take Webster "a *long time* to acquire Mr. Lucas['s] scientific knowledge," but he considered Webster "a treasure, a jewel of a man" and advised Ward to pay him a trial salary that would "enable him to marry at once and settle contentedly down." Indeed, Webster did marry in 1882 and stayed with Ward for the remainder of that year, but then he, too, moved to Washington. However, he chose not to join his former colleagues at the National Museum; instead, he established a private taxidermy enterprise where he had the prospect of greater income and autonomy.

Webster viewed taxidermy as divided into two groups, "Museum or Institutional Taxidermy," which he determined was "conservative and settled," and "Decorative Taxidermy," which he argued was "radical and progressive."[4] In his 1883 lecture to the Society of American Taxidermists, titled "Taxidermy as a Decorative Art," he described the "museum conservative" as a taxidermist who may have "artistic skill," but "whose position or class of work demands from him the stiff or contracted style of work the average institution demands—mark you, demands."[5] While he conceded that these demands were reasonable, he argued that "too often the strict commission comes to the taxidermist: No fine work, no fancy positions—we want plain, straight styles. That simply means a lot of specimens of the 'straight-jacket' order."[6] For Webster, the preparation of scientific specimens—taxidermy's historical connection to the museum institution—was the very reason why taxidermists had struggled unsuccessfully to advance their art. He described the taxidermist's relationship to the museum as "pernicious" and likened it to "keeping a blooded horse carting stones before a fifteen-hundred pound cart, and expecting him to develop into a trotter."[7]

Decorative taxidermy, for Webster, was the antidote; his Washington studio letterhead described him as an "Artistic Taxidermist." Though he mounted groups of birds and mammals for the U.S. National Museum—including a striking group of five African *Colobus* monkeys climbing on a leafy bough that was praised by the *Washington Post* as "extremely artistic and lifelike"—the majority of his fame came from doing simple single-specimen mounts for high-profile clients, such as Stonewall Jackson's horse for the general's widow and a deer for Grover Cleveland mounted as a "memento" of the president's hunting trip to West Virginia.[8] Webster also advertised his skill in preparing feather fire screens, bird medallions, game panels, rugs, robes, horn and antler furniture, and "bric-abrac novelties."

By contrast, Carl Akeley's Milwaukee studio letterhead announced his dual expertise in "Scientific & Decorative Taxidermy" and emphasized scientific taxidermy for "museums, colleges, etc.," with a "specialty of fine group work." Where Webster argued for the acceptance of decorative taxidermy in natural history museums, Akeley focused on how to appropriately unite decorative or artistic taxidermy with science. Akeley believed that scientific taxidermy required measurements of the animal taken in the field directly after death, knowledge of anatomy, and a study of the animal in its habitat. With these scientific tools, the taxidermist could then apply recently developed artistic methods, including the techniques of "manikin making," clay modeling, and design of accessories such as leaves and branches, in creating an accurate representation of the animal and its habitat.[9]

While at Ward's, Akeley, like Hornaday, was discouraged by the application of the "old straw-rag-and-bone method."[10] So once he learned animal anatomy, he asked Ward if he could experiment with a new mounting method on a zebra recently acquired by the establishment. Ward agreed, but characteristically cautioned Akeley that he could work on the project only after hours. Akeley made "a plaster cast of the body," believing that it would facilitate the making of an anatomically correct mount. Despite these efforts, Ward deemed the method too time-consuming and too expensive to adopt at the establishment. Akeley later opined that "the zebra was handed out to be mounted in the old way and my casts were thrown on the dump."[11]

Frustrated by the limitations at Ward's, Akeley left Rochester in November 1886 to join his close friend William Morton Wheeler, another former Ward's employee, in Milwaukee. The Milwaukee Public Museum had recently opened its doors, and Wheeler saw an opportunity for Akeley to practice his new methods in preparing mounted specimens for the muse-

Fig. 4.1. Carl E. Akeley in Milwaukee, circa 1886. (Image #212489, American Museum of
Natural History Library)

um's exhibits. Akeley set up his scientific taxidermy studio in a barn on the
Wheeler family's property and immediately set to work. After Wheeler was
appointed custodian of the museum, he created a position for Akeley as
the museum's taxidermist—a half-time post at first, then, in July 1889, ex-
panding to full-time.[12]

Though Wheeler was technically Akeley's boss, he revered Akeley's
taxidermy work. "We were nearly of the same physical age," Wheeler remembered later, "but I was the younger and more unsettled mentally." He
admired Akeley's work ethic, his quiet sense of humor, and his "thoroughly
manly disposition."[13] More than a supervisor, Wheeler was an acolyte and
apprentice, and he granted Akeley great latitude in his work. In turn, Akeley used this freedom to develop, step-by-step, a new method for mounting
large mammals—both at the museum shop and in his private studio. In the
evenings, while Akeley mounted specimens, Wheeler read "a whole small
library" aloud to him. Wheeler fondly recalled, "Perhaps Akeley really heard
only occasional important fragments and had found that he could carry on
his own trains of inventive thought better when we were together and I was

making a continual but not too disturbing noise."[14] In those hours, Akeley made use of Hornaday's method of covering a manikin in clay, but improved on the technique by using field measurements and, if possible, photographs and anatomical casts of the specimen taken just after death. This method resulted in "a model not only of the species but of the actual animal whose skin we were going to use."[15]

Akeley also recognized that this process should not be the final stage before mounting the specimen, as there were problems associated with the use of clay under skin—particularly humidity, which caused the skin to expand and contract and eventually to crack. After much experimentation, Akeley concluded that "a papier-mache manikin reinforced by wire cloth and coated with shellac would be tough, strong, durable, and impervious to moisture." After making a plaster mold of the clay model, Akeley coated the inside with glue, on which he placed "a sheet of muslin and worked it carefully and painstakingly into every undulation of the mould." Next, Akeley pressed "thin layers of papier-mache" (the number of layers depended on the size of the specimen) with "wire cloth reinforcement" into the mold. To make the manikin impermeable to water, he coated each layer of the papier-mâché with shellac. Once it had dried, Akeley submerged the mold in water, which "affected nothing but the thin coating of glue between the mold and the muslin. That melted and my muslin-covered, reinforced papier-mâché sections of the manikin came out of the plaster mould clean and perfect replicas of the original clay model."[16]

In an effort to showcase his new method, Akeley proposed to mount "a series of groups of the fur-bearing animals of Wisconsin, the muskrat group to be the first of the series." Among the moneymen on the museum's board, the idea, Akeley remembered, was "more tolerated than encouraged," but with Wheeler's influence, the exhibit was approved. One year later, in 1890, Akeley completed the experimental group, which remains on display at the museum to this day. It depicts a family of five muskrats preparing their domed nest of grass, roots, and mud in the marshy shallows of a Wisconsin lake as winter approaches. Akeley's innovative design revealed a cross-section of the marsh, above and below the waterline, as well as the inner tunnels and chambers of the nest. Wheeler praised Akeley's ability to overcome "the great difficulties in accurately imitating the boggy earth, the half dead vegetation and the stagnant water."[17] Soon after the exhibit opened, however, Wheeler, finding that his position as museum custodian afforded him little time to pursue his growing research interests in entomology, resigned and accepted a fellowship at Clark University in Worcester, Massachusetts.[18] Without Wheeler's support, the board declined to fund similar dioramas. The muskrat group was the only exhibit completed in the Wisconsin mammal series. Akeley remained on for two more years, occasionally experimenting with casting and mounting large mammals—most notably a group of orangutans—but he chafed at the lack of funding for his ambitions.

On September 20, 1892, Akeley left the museum to pursue his private scientific taxidermy business. In the same year, Wheeler earned a PhD from Clark and was appointed instructor in embryology at the University of Chicago. Wheeler aptly characterized himself and Akeley as of a certain type who were "probably endowed with a more unstable if not more vivid imagination," with a "subconscious dread of being owned by people and things and soon exhaust the possibilities of their medium, like fungi that burn out their substratum, and become dissatisfied and restless till they can implant themselves in fresh conditions of growth." Knowing that Akeley was eager to design larger exhibits using his new scientific taxidermy methods for a world-class natural history museum, Wheeler returned in 1895 from a research trip to England with an offer from Sir William Henry Flower, director of the British Museum of Natural History, for Akeley to lead its preparations staff.[19]

AKELEY AT THE FIELD MUSEUM OF NATURAL HISTORY

Before accepting Flower's offer, Akeley decided to visit Wheeler in Chicago and see the Field Museum of Natural History (then known as the Field Columbian Museum). While there, Akeley met with Daniel Giraud Elliot, the museum's curator of zoology. Elliot was impressed by Akeley's new taxidermy methods and suggested that instead of accepting the British Museum offer, he should stay in Chicago and serve as the Field Museum's chief taxidermist. Elliot also invited Akeley to accompany him the following year on a collecting expedition to Africa.[20] Akeley, who couldn't resist the lure of another field expedition, accepted his offer. On May 11, 1895, *Forest and Stream* reported that "Mr. Akeley is this summer to come to Chicago as taxidermist for the Field Columbian Museum, where his really artistic work will no doubt be admired by many." The magazine's announcement also heralded Akeley's new method: "In his process of modeling all of the art of the sculptor seems in evidence. . . . There is a new feature in taxidermy, that of a conception, a modeling, so that one must call the workman not merely a workman, but also an artist."[21]

In March 1896, Akeley and Elliot departed for Africa. Elliot, an advocate for international wildlife conservation, cabled to the press that the "rapid disappearance of wild creatures in Africa made it necessary for the expedition to go upon the field before it was too late."[22] News of the trip received national coverage. The *New York Times* reported that the expedition would focus on collecting "those species which are becoming extinct. The ordinary way of securing such specimens—one at a time—from collectors is slow and uncertain, and for this reason the directors of the institution determined to send Prof. Elliot directly to the country."[23] The conservation-minded tone of the trip highlights the fact that, despite Akeley's maverick approach to taxidermy, he was conscious of the efforts of his esteemed predecessors Lucas and Hornaday, who believed that the mission of the new taxidermy movement must include the work of saving endangered wildlife at home and abroad.

Even before the Field Museum, the Carnegie Museum, and the American Museum had launched scientific expeditions to Africa at the turn of the twentieth century, the continent was a source of enthrallment for the American public. Huge crowds turned out for the international lecture tour of journalist and explorer Henry Morton Stanley (famous for his search and rescue of missing missionary David Livingstone); books by African explorers, including Paul Du Chaillu, became best sellers; and many Evangelicals in America viewed Africa as "the laboratory of Christianity."[24] In direct response to this rising interest, directors of major metropolitan natural history museums, looking to expand their public exhibits, began planning elaborate African halls. Institution-sponsored scientific collecting expeditions, often funded by wealthy board members or their friends, were by now the means by which most museums acquired specimens, as they had adopted, almost universally, "a policy of exploration," instead of purchasing specimens from suppliers like Ward's Natural Science Establishment—recognizing the trend, Ward's had already shifted its focus toward educational series of specimens for classroom use.[25]

Specimens collected on these expeditions were not, however, solely for exhibition. Scientific series were also collected to support the research of American scientists, who had a more focused interest in Africa and its fauna. Elliot was one of a group of scientists concerned with the world's rapidly disappearing mammalian fauna, including Wilfred H. Osgood, also of the FMNH, Charles C. Adams of the Roosevelt Wild Life Experiment Station, and Edward W. Nelson, chief of the Biological Survey. This group agreed with Henry Fairfield Osborn's theory "that the predominant fauna of America in the Middle and Upper Miocene Age and in the Pliocene was closely analogous to the still extant fauna of Africa," and that a careful study of African species might help avert what Osborn would later term the "Close of the Age of Mammals."[26]

Although institution-sponsored expeditions were now the main source of scientific and exhibit specimens for American natural history museums, there were still exceptions. While Akeley and Elliot were in Africa, the Ringling Brothers Circus donated the carcass of a giraffe to the FMNH. According to the *Chicago Tribune*, "After being cured and prepared the skin is to be left until the return of Mr. Akeley, the taxidermist of the museum, who is now in Central Africa. When he gets back it will be stuffed in a natural position."[27] Even as the *Tribune* excitedly noted that a giraffe would soon be on display at the Field Museum, it also acknowledged the looming threat of large mammal extinction—albeit blaming Africans for the decline. The article noted that Akeley would "undoubtedly attempt" to secure a giraffe, but there was "such a scarceness of the commodity that considerable doubt is expressed as to his success. . . . Now that Africa has ceased to keep up the supply it is probable that a few years will see nothing but stuffed giraffes on exhibition."[28]

On November 22, the *Chicago Daily Tribune* reported that Elliot and Akeley had reached St. Louis on their return from Africa, having secured "the finest collection of large mammalia ever obtained in a single expedition—nearly everything they went after."[29] News of the expedition quickly established Akeley's heroic reputation.

"Were you ever in fear of wild beasts?" the *Tribune* reporter asked.

"Mr. Akeley could tell you of an exciting adventure he had with one," Elliot replied.

While hunting at dusk for specimens, they said, Akeley had come upon a female leopard. He fired, but succeeded only in wounding it in the right hind foot. The leopard attacked Akeley, knocking the rifle from his hands. Lunging for his throat but missing, it managed to bite down on his upper right arm. Akeley fought the leopard by kneeling hard against its chest and breaking ribs, as he simultaneously pushed its cheek in until the animal was biting down on the inside of its own mouth.[30]

"The leopard released the hold on his arm," the reporter recounted to readers, "and soon with knees and hand he had the beast's life choked out."[31] Two of Akeley's fingers were badly chewed, and he had fourteen wounds in his arm and shoulder. He made it back to the camp, where Professor Elliot disinfected his wounds. Akeley made a full recovery—with a story to tell. For decades to follow, he told and retold, in interviews and in writing, of his near-death encounter with the leopard. It became a central part of the

Fig. 4.2. Carl E. Akeley was injured by a leopard while collecting for the Field Museum in Africa in 1896. (Courtesy, Field Museum. Image #CSZ5974)

Akeley myth: the museum taxidermist willing to risk everything to collect specimens for science and for his dioramas.

By late December, under Akeley's supervision, the museum staff, including William Henry Holmes, curator of anthropology, and his assistants, joined in unpacking the "carload" of material packed in crates and barrels of brine:

Among the most valuable trophies are the skins of a great many African animals, birds, and reptiles. These are mostly of rare species, some of which are nearly extinct. There are skins of lions and leopards, of the oryx, baboon, jackal, hyena, hedgehog, and numerous rare specimens of African antelope and gazelle. Among them the most highly prized specimens obtained are of the Dik Dik and Koodoo antelopes and Saemerring, the Peizean, the Clarke, and the Waller gazelle.[32]

Under Elliot's supervision, Akeley immediately began planning groups and mounting specimens.

In his annual report to the Field Columbian Museum Board of Trustees, Director Frederick J. V. Skiff praised Akeley's first three groups: "The striking manner in which these three rare and interesting groups of animals are arranged and posed, the life action and naturalness of the picture presented, no less than the scientific fidelity and faithfulness of accessories, stamp them at once as of the very highest character of work that can be performed."[33]

The group of lesser kudu and that of Waller's gazelles depicted arid scenes of the Somali desert. The lesser kudu family group had six representative specimens as well as an African owl species perched atop an anthill, from which grows a small tree with vine-like branches and leaves, on which one of the kudu feeds. "Although the area is limited," Skiff noted, "the impressions of the desert are forcibly conveyed to the spectator."[34] The family group of six Waller's gazelles, or gerenuk (Somali for giraffe-necked), Skiff said, "presents that graceful animal in [a] most effective and dramatic grouping, finished in every artistic detail, and complete in every requirement of the scientist and hunter."[35] The most impressive feature, Skiff believed, was the male standing on its hind legs with a foreleg leaning against a tree trunk as it stretches its neck to feed on the green leaves of a tall branch—common behavior among the species. Two watchful animals, one adult male and a yearling, stand in the background, and a female chews on a leaf as two young lie hiding among succulents.

At about the same time, Akeley also completed a family group of musk oxen mounted for the North American mammals exhibit, which Skiff described as "full of quiet, natural life; everything is harmonious and realistic." The group comprised seven animals gathered on a "field of snow, through which a huge rock protrudes, surmounted by a splendid male in a commanding attitude."[36] In the foreground, one animal digs in the snow "with hoof and nose for what lichens may be concealed beneath."[37]

The groups were placed in large mahogany plate glass cases and installed in the museum's west court early in 1898. Before the official exhibit opening

Fig. 4.3. Carl E. Akeley's group of Waller's gazelles mounted for the Field Museum.
(Courtesy, Field Museum. Image #CSZ12557)

at the museum, *Science* reported, "The mammal groups of Mr. Akeley, who is unrivaled in this work, are deserving of special notice, particularly that of the Lesser Koodoo with its striking central figure. The group of Musk oxen contains, so far as we are aware, by far the best series contained in any museum."[38]

Akeley was able to achieve this success largely through the unprecedented support he received from Elliot, who lobbied the museum's administration on his behalf. In 1898, Elliot appealed to Skiff to provide Akeley with improved working conditions and assistants to augment his work; he also submitted a report indicating how best to expand the taxidermy division, outlining what materials would be required. As the exhibit collections grew, they needed to be prepared and stored until they could be mounted—space limitations were common to all museum preparations departments of this period.[39] As long as Akeley continued to create outstanding mounts and positive publicity for the museum, Skiff honored Elliot's requests. One year later, a second story was added to the taxidermist's shop.[40]

When Akeley wasn't mounting African specimens, he and his wife, De-lia J. Akeley, worked in their private workshop on a quartet of Virginia deer groups called "The Four Seasons." Akeley had conceived of this ambitious exhibit while at the Milwaukee Public Museum and had intended to sell it to that museum.[41] The exhibit depicted a family of four deer—a buck, a doe, a yearling "spike" buck, and a fawn. Beginning with the "Summer" group, the visitor could follow the family through the year, viewing changes in seasonal coats, the growth of the buck's antlers, and the growth of the fawn from ten days old to a yearling. The exhibit's construction was as innova-tive as its content. Akeley divided one large case into four sections, and—to keep the viewer from seeing the other groups through the glass in the back of each case—he installed behind each group a curved background on which was painted, by C. A. Corwin of the Chicago Art Institute, a scene of the appropriate season.

Akeley hoped to convince Elliot to acquire the exhibit for the FMNH. But Elliot said he couldn't justify the expense of buying all four groups; in-stead, Akeley may have used the deer groups to persuade Elliot to let him design a hall of North American ruminants—a long-term plan that could be funded over time.[42] The two began planning for the new hall, and in July 1898, they traveled to the Pacific Northwest to collect more specimens of North American mammals. The rugged and unknown terrain made for slow travel, but Elliot reported in September that they had collected "five hundred skins of deer, carnivora, and rodents."[43] One month later, the ex-pedition returned to Chicago with numerous specimens, including a new species of mouse that Elliot named for Akeley, *Peromyscus akeleyi*.[44]

Back from the field, Akeley mounted additional African specimens, in-cluding a cheetah group and a group of striped hyenas. Each of these groups consisted of several individuals feeding. The family of cheetahs is gath-ered around a Soemmerring's gazelle—the alert adults with their attention turned from the kill as the cubs wander around the gazelle—while the group of hyenas depicts four adults feeding on carrion. Though he did not depict these carnivores in the tranquil domestic scenes advocated by Hornaday, he nevertheless departed from the fighting poses of expositions and early museums.

While Akeley was completing the African series, the zoology depart-ment expanded its exhibition space into the museum's south court. In this new space, Akeley installed a group of polar bears. The exhibit depicted a male coming upon a female with a yearling and an older cub feeding on a ringed seal. The female aggressively warns the male away, knowing that he might kill the yearling. Akeley's depiction of the male not as family

Fig. 4.4. Carl E. Akeley's group of cheetahs mounted for the Field Museum.
(Courtesy, Field Museum. Image #CSZ62835)

Fig. 4.5. Carl E. Akeley's group of striped hyenas mounted for the Field Museum.
(Courtesy, Field Museum. Image #CSZ62842)

Fig. 4.6. Carl E. Akeley's polar bear group mounted for the Field Museum. (Courtesy, Field Museum. Image #CSZ6243)

sentinel, but as intruder, was based on the latest scientific research.[45] The input of an eminent zoologist like Elliot allowed Akeley ready access to valuable information about animal behavior and habitat, which—coupled with his superb grasp of animal anatomy—allowed Akeley to make groups that were not only artistically appealing but also revolutionary in their accuracy.

For Akeley, the museum atmosphere was supportive and even stimulating, but for Webster, who did not have mentors in the scientific community, the museum environment remained a hurdle to creativity.

WEBSTER AT THE CARNEGIE MUSEUM OF NATURAL HISTORY

For a time, working outside of the museum institution allowed Webster the freedom to choose his work and further develop his art. Yet he maintained the U.S. National Museum as his principal customer, for which he mounted numerous specimens of single birds and bird groups, as well as small mammals.[46] Soon, however, he found there were constraints in maintaining a private enterprise. Financial concerns forced him to take jobs that were not always opportunities for artistic achievement, including the mounting of pets—a popular Victorian custom—to keep the business afloat.[47]

In 1892, he moved to New York City as a partner in Sowdon & Webster Furriers and Taxidermists on Broadway. The firm quickly dissolved, and Webster opened his own studio on Fifth Avenue, but that, too, failed. By 1895, he had moved to Mount Vernon, New York, and had become the secretary of the First Sportsman Association Exposition, held at Madison Square Garden—reduced to depicting "pleasant" hunting scenes.

So when the offer came in 1897 to join the staff of the new Carnegie Museum of Natural History, it was financial difficulties, rather than high ideals, that drove Webster to join the ranks of museum taxidermists. Nevertheless, he must have enjoyed the freedom of his position as a CMNH "preparator"—a job that included both taxidermy and osteology. In its first annual report, the museum's director, W. J. Holland, announced that Webster was "one of the ablest and most widely known artists of his class,"[48] whom he had chosen for "his talents in the construction of groups which shall illustrate the life-history of various important classes of animals."[49] The museum administration had already adopted a standard design from the National Museum for its exhibit cases, but Webster was given great latitude in designing the groups that would fill them.

In 1898, the CMNH opened two exhibits—the museum's first habitat groups. The first, a flamingo group, was an elaboration of Webster's infamous "The Flamingo at Home," but with its inaccuracies corrected. The group's four males and three females were mounted in various attitudes, with one bird sitting with legs folded atop its nest, the way Webster had intended to mount the incubating bird in the original group. The second was a group titled "California Condors and Turkey Buzzards on Dead Wapiti."[50]

Of the latter group, Holland boasted that it "challenges comparison with anything of like character in the museums of this country, or Europe."[51] The uniqueness of this exhibit lay in the fact that Webster successfully incorporated for the first time an artifact from an invisible third participant—a Native American hunter, who is represented only by the inclusion of an arrow protruding from the wapiti's side. The scene was meant to conjure the past of perhaps one or two hundred years earlier, when these species were abundant in North America. After being shot with an arrow, the elk has managed to elude the hunter, but not death. Yet the dead animal is not wasted, as the California condors and the turkey vultures are feasting on the meat. The conservation message focuses on two threatened species: the Tule elk, once

Fig. 4.7. Frederic S. Webster's "California Condors and Turkey Buzzards on Dead Wapiti," mounted for the Carnegie Museum of Natural History in 1898. The group is still on display in the museum. Source: *The Carnegie Museum, Annual Report of the Director for the Year Ending March 31, 1898*, facing 65. (Carnegie Museum of Natural History)

a thriving herd of over a million, but by then reduced to only twenty-eight individuals,[52] and the California condor, already near extinction.[53]

Next, Webster showed his commercial taxidermy sensibilities by mounting the hunting dog "Count Noble" and by acquiring and remounting Verreaux's famous exposition piece, "Arab Courier Attacked by Lions." Both were unveiled in November 1899.[54] The "Count Noble" group, installed in the museum's bird hall, featured an English setter flushing a covey of quail.[55] Near this group, Webster installed smaller exhibits of cherrybirds in the branch of a cherry tree, an American robin on the branch of a blossoming apple tree, a Baltimore oriole in the branch of an elm tree, a golden-winged warbler with a nest in a clump of weeds, and a chestnut-sided warbler with a nest in a blackberry bush. All of these exhibit cases included birds with nests and eggs.[56]

"Arab Courier" was obtained through exchange. Joel A. Allen, curator of ornithology and mammalogy at the AMNH, "realized [the exhibit's] historical value as well as its enormous potential to attract and delight visitors, and he knew that Andrew Carnegie's new museum in Pittsburgh was in need of impressive new exhibits."[57] The AMNH administration probably chose to

Fig. 4.8. Habitat groups mounted by Frederic S. Webster from 1898 to 1900 for the Carnegie Museum of Natural History: "Count Noble" group (*left, foreground*), "California Condors and Turkey Buzzards on Dead Wapiti" (*center*), and a group of brown pelicans (*right*), circa 1911, after the Carnegie Museum of Natural History had moved to its new building. (Author's collection)

donate "Arab Courier" because they preferred more tranquil domestic family groups. When the group arrived in Pittsburgh, Webster found it in poor condition. Because it had never been displayed in a glass case, it had been exposed to light and variations in humidity, which caused the fur and the natural fibers used in the accessories to fade and the skin to crack. Before reassembling the various pieces, Webster began an intensive restoration project:

> Wood, wire and excelsior were added to the [camel's] neck for additional support and the ears and eyes were repaired. He also remodeled the mouths and tongues of the camel and lions and cleaned and waxed their teeth. In addition, the hides, clothing, saddle and other paraphernalia were thoroughly cleaned.[58]

Once completed, the group was placed in a glass case and was re-presented to the public in November 1899. Holland reported that the "group proves very attractive, especially to younger visitors, and through Mr. Webster's skillful manipulation it is in even better condition than when originally set up by the taxidermist."[59]

In spring 1899, while still working on these two groups, Webster went on a field expedition to Brevard County, Florida. He collected widely on the trip, bringing back specimens of an American crocodile, alligator, several species of snakes, and "an extensive series of land-shells." However, his primary mission was to gather "the material for an important group of Pelicans"—not only the specimens, but also the accessory material.[60] Even Webster was coming around to accepting Akeley's concept of museum taxidermy as necessarily including both scientific and artistic qualities, as he found it important to collect the bird's nests, as well as vegetation and driftwood from the same locality, to complete the design of his pelican group.

Frank M. Chapman, ornithologist at the AMNH, had visited Pelican Island and other pelican rookeries along the eastern shore of Florida in 1898; he presented a paper on the nesting habits of the brown pelican later that

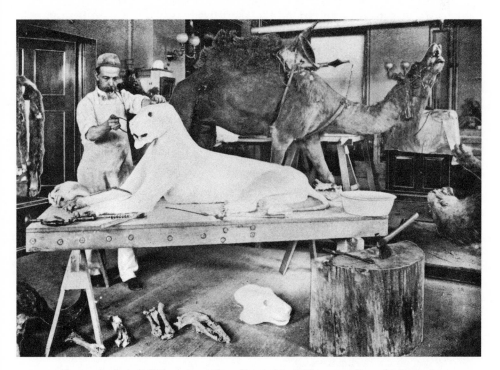

Fig. 4.9. Frederic S. Webster mounting a lion in his taxidermy studio at the Carnegie Museum. Note the dismounted "Arab Courier" in the background. Source: *The Carnegie Museum, Annual Report of the Director for the Year Ending March 1900.* (Carnegie Museum of Natural History)

year at the annual congress of the American Ornithologists' Union (AOU).[61] Chapman had made a similar trip to Bird Rock Island in the Gulf of St. Lawrence, near Nova Scotia, that same year, where he had amassed an extensive collection of specimens and photographs for "The Bird Rock Group" at the AMNH. Former Ward's taxidermist Harry C. Henslow was already at work on the impressive group, which would feature a realistic rock face and more than forty seabirds representing seven nesting species: the razor-billed auk, petrel, gannet, puffin, kittiwake, common murre, and Brünnich's murre.[62] Webster may have gone to Florida to collect specimens in order to compete with the AMNH, but the trip was his first close-up glimpse of the devastating effects of overhunting by feather collectors—and it seems to have brought about a change of heart.

Fifteen years earlier, Webster had spoken before the first congress of the AOU of his concerns about the work of its bird protection committee, which he feared would prove too restrictive and might threaten "to prevent work in legitimate taxidermy."[63] In 1891, Webster had gone so far as to found a rival organization—the short-lived Association of American Ornithologists—which held its organizational meeting in his Washington taxidermy studio. Now, however, working within the context of a scientific museum and newly armed with firsthand knowledge of the devastation wrought by the millinery industry in its quest for feathers, Webster returned to Pittsburgh eager to mount the nation's first group of pelicans, "representing both adult and young birds in all stages of plumage,"[64] in an effort to bring their plight to the attention of museum visitors. In August, the *Pittsburgh Press* reported that Webster was in the process of preparing five bird groups:

> The large group will be made of large, odd-looking birds of a species, which is never seen in these regions. The name and nature of it is to be kept a secret until founder's day. Each of these groups will show the bird in its natural environments, also the manner of life and breeding habits of the birds. They will be made as nearly natural to life as it is possible to make them.[65]

But Webster ran into delays in completing the group, and it was not until the following Founders' Day, November 4, 1900, that he finally installed the "secret" brown pelican group.

It was the largest and most ambitious of Webster's bird groups. The ground replicated the Florida shoreline, littered with driftwood, on which the birds had built several nests. One nest held a recent hatchling, another held four eggs, and the others had adults sitting upon them. In one corner of the case, five young birds fed from the beak of a female, while on the opposite side, a large male perched watchfully on a bare, low-hanging branch. Another adult hung from invisible threads, about to land on the beach.[66] The exhibit had a powerful and immediate effect. A few months after the group was installed, Florida legislators—at the urging of the AOU and the Florida Audubon Society—passed a law to protect non-game birds from market hunters. After two of the four wardens hired to enforce the law were murdered, Paul Kroegel, one of the two remaining wardens, convinced Chapman that the only way to fully protect the last brown pelican rookery on Florida's eastern coast was to enact federal legislation to set the land aside. After a meeting with Chapman in March 1903, President Theodore Roosevelt issued an executive order establishing Pelican Island as the first federal bird reservation—the precursor of the National Wildlife Refuge System.[67]

In 1901, Webster unveiled another new group of a Florida bird species, the ivory-billed woodpecker. This small group contained only two individuals, one male with its characteristic red crest and one black-crested female, searching the bark near their nest hole for wood-boring insects. Comparatively simple in execution, the mount focused not on the taxidermy, but on the content of the accompanying exhibit label. The placard referred to the "tragic history" of the species and described it as "on the verge of extinction."[68] Less than a decade earlier, some ornithologists had argued that the ivory-billed woodpecker would not be threatened with extermination because, as one writer put it, "Surely a bird as wild, as wary, would not remain in an area where man was constantly to be met!"[69] Here, however, Webster took the opportunity to better educate the public. On the exhibit label, Webster emphasized that the ivory-billed woodpecker was not the victim of "DIRECT persecution by man," as were the brown pelicans; instead, they were rapidly disappearing because their native swamps "have been so extensively logged for their valuable timber that the type of forest required by the woodpeckers is almost entirely gone."

Webster had undertaken an ambitious project to translate a sophisticated scientific concept for the public: the idea that a species may be endangered not as the direct result of killing the animals, but by the destruction of its habitat and food source. Webster had finally discovered the educational value of his decorative masterpieces—works that could not only delight aesthetically, but also inform and perhaps even sway public opinion. Only by bringing together the latest scientific research and the best decorative methods could habitat groups serve as adequate educational tools. Webster's idea now converged with Akeley's, rather than competing with it.

Ironically, Webster had set a standard that he himself would soon fail to meet. Even as he was achieving new heights in his public displays, Webster was allowing the scientific collections under his care to deteriorate. Though he now understood the value of decorative taxidermy in educating the broader community, he failed to recognize the value of well-preserved specimens to the scientific community. He had never favored rows of "plain, straight . . . specimens of the 'straight-jacket' order"[70] for public display, and he certainly did not see the importance of giving his limited time and scarce resources to collections that would be viewed by only a select few individuals.

AKELEY AND "THE FOUR SEASONS"

In January 1900, FMNH director Frederick Skiff, in a rare show of support for Akeley's expensive exhibit work, requested that Harlow N. Higinbotham, president of the FMNH Board of Trustees, purchase all four groups in Akeley's "The Four Seasons" for the museum for the sum of $5,500, explaining that Akeley had "started to make this his piece de resistance, and undiscouraged . . . will undoubtedly present a work of great excellence."[71] Akeley had devoted four years to mounting the four groups. In the end, he determined that financially he had made no profit; although he had broken even on the cost of materials and labor, he had not been paid for his time.[72] In retrospect, he felt that "it was a pretty good four years' work" because, in the end, he "had the experience and the method."[73]

Two years later, the exhibit opened at the FMNH. It was an immediate sensation, covered in the Chicago newspapers and universally praised and embraced by museum professionals. On August 24, 1902, the *Chicago Daily Tribune* announced that "The Four Seasons"—"an innovation in the mounting of groups of animals"—had recently opened at the Field Museum. "Each of the four groups is set in the most perfect of woodland scenery," wrote the *Tribune* reporter:

> The summer scene is the edge of a swamp in the midst of the dense woods. The autumn study is mounted in a "burning" with fire eaten stumps left standing and the ground strewn with dead leaves. The winter scene shows the ground thickly carpeted with snow, through which the deer are tracking about. The spring picture is set in an open wood, with the trees hardly in bud and the ground covered with moss.[74]

Even before the exhibit opened, Frederic Lucas, now in charge of the Department of Biology exhibits at the U.S. National Museum, wrote Akeley

Fig. 4.10. Carl E. Akeley's "Summer" group of Virginia deer from "The Four Seasons."
(Courtesy, Field Museum. Image #CSZ6212)

requesting copies of local newspaper accounts of "The Four Seasons." As he explained, "I have the misfortune to be the American correspondent of the Museum's Journal of Great Britain and should be glad to send a description of the group to the editor, particularly as he is rather mournful over the small amount of news he obtains from the United States."[75] Akeley sent Lucas photographs of "The Four Seasons," and by May of the following year, *Science* had published Lucas's description and praise of the groups: "These have been in preparation for a long time past, and are unquestionably the most elaborate of the kind anywhere, and the most successful of attempts to imitate nature in museums."[76]

On August 2, 1903, Akeley received a letter from his friend William Alanson Bryan, who had left his position as assistant curator of ornithology at the FMNH to become curator of ethnology and natural history at the Bishop Museum in Honolulu. Bryan, after seeing the Field Museum's annual report for 1902, which contained photographs of "The Four Seasons,"

Fig. 4.11. Carl E. Akeley's "Autumn" group of Virginia deer from "The Four Seasons."
(Courtesy, Field Museum. Image #CSZ6208)

was "delighted with them" and joked that the "only disparaging thing that
I could offer would be a complaint—for you have left little indeed for your
colleagues and successors to accomplish in the way of perfecting the technique of the art of representing nature [but] nature itself."[77] He went on
to claim that Akeley had "done for Taxidermy what Michelangelo did for
sculpture and painting."[78] He also took the opportunity to caution his friend
to publish his new method "before it is too late—you owe it to *Akeley* and
to Taxidermy."[79]

Two years earlier, in January 1901, Hornaday wrote to Akeley, praising
his accomplishments: "In thinking over your surprisingly fine groups of
mammals, I am more and more strongly impressed by their artistic excellence. They are the finest examples of taxidermy in the world—so far as I
have seen it."[80] The *New York Herald* had commissioned Hornaday to write
a "special article" on modern European taxidermy. But Hornaday intended

to "sidetrack" the focus and write instead "a pan-American article." He therefore requested that Akeley send him photographs of his groups so that he could include one in the article.

Hornaday was most impressed with a group of greater kudu for technical rather than artistic reasons, stating in a letter to Akeley that the group "shows off your work to better advantage than any of the others, because of the closeness of their hair."[81] Hornaday ultimately used photographs of three of Akeley's groups—the greater kudu group, a Swayne's hartebeest group, and a Somali wild ass group—as well as an image of a single Stone's ram in the *Herald* article.

Hornaday's article featured Akeley's work more than that of any other taxidermist, including Webster, whose "California Condors and Turkey Buzzards on Dead Wapiti" was the only other group he mentioned. Hornaday admitted that he had not traveled to Pittsburgh to visit the CMNH, so he

Fig. 4.12. Akeley's "Winter" group of Virginia deer from "The Four Seasons." (Courtesy, Field Museum. Image #CSZ6213)

Fig. 4.13. Akeley's "Spring" group of Virginia deer from "The Four Seasons." (Courtesy, Field Museum. Image #CSZ62847)

hadn't even seen the group—though he later included it in his "Master-pieces of American Taxidermy." For inclusion in the *Herald* article, Hor-naday also used single photographs representative of the work of Lewis L. Dyche, at the University of Kansas; William Palmer, at the U.S. National Museum; and John Rowley Jr., at the AMNH. From this younger genera-tion of taxidermists, Akeley was singled out for highest praise. Hornaday described Akeley's work as "genius," and "something for every American to be proud of."[82] Hornaday went on to argue that "there is no taxidermic work in Europe which equals" that of Akeley. He elaborated:

> When I saw Mr. Akeley's splendid array of groups of African mammals—close haired, every one of them, exquisitely modeled and posed, neither fat nor lean and not a seam visible anywhere—it gave me a thrill of plea-sure to think that those masterpieces of taxidermy had been produced in America.[83]

As Hornaday was not known for his humility, Harry Denslow, Akeley's former assistant at the FMNH, was surprised at his public approval of Akeley's taxidermy, writing, "Though he certainly gave credit to some old friends more than their due. I was glad to see he spoke without prejudice against your work."[84] Hornaday—who had once referred to himself as the nation's leading taxidermist of both small and large mammals—no longer competed for the title of best taxidermist. As director of the New York Zoological Park, he was content to stand outside the fray, as long as he was recognized as the founder of the new American taxidermy.[85]

Once the article was published, Akeley wrote to Hornaday, expressing his appreciation. Hornaday responded by praising again Akeley's "really wonderful work" and encouraging him to continue: "I trust that you will never grow tired of taxidermy, but that as years go by your interest in it will increase as it evidently is doing now."[86] When Akeley informed Hornaday that he intended to publish his new taxidermic method for mounting large mammals, Hornaday responded enthusiastically: "I am glad you are going to publish your new method in the mounting of mammals, and hope that it will be brought out by the museum in fine shape, regardless of expense, illustrated by many examples of your work."[87] But the planned publication never materialized.

With all of the positive attention, Akeley began "to dream of museums which would have artist-naturalists who would have the vision to plan groups and the skill to model them and who would be furnished with skilled assistance in the making of the manikins and accessories and in the mounting of the animals."[88] He believed that the dream was close to being realized when, in 1904, he met with Hermon C. Bumpus, director of the AMNH, who told him that he had recently hired James Clark, "who could model but who did not know the technique of making manikins and mounting animals."[89] Akeley agreed to let Clark study with him in Chicago and learn his new method. Together, the two mounted a Virginia doe. Clark took the mounted specimen back to the AMNH, where it was placed on display.

Shortly after Clark returned to New York and the Virginia doe was installed at the museum, *Scientific American* ran an article declaring the taxidermy department at the AMNH to have recently "adopted what is considered the most advanced and artistic method of reproducing wild animal life for exhibition purposes so far devised. In fact, it is almost revolutionary in its way, since it involved a radical departure from the old, stereotyped rules of taxidermy."[90] The reporter talked to Clark, who described Akeley's method in extreme detail—but never credited his instructor. Akeley was incensed.

About this time, William A. Bryan visited Akeley in Chicago. Bryan, on behalf of the Bishop Museum, was making an extended trip, touring all the major American natural history museums, and volunteered to gather intelligence for Akeley when he was at the AMNH. A week later, Bryan followed through, delivering disturbing news. He had spent more than four hours with Clark in the taxidermy laboratory, where he saw a group of elk in progress that had been highly praised in the *Scientific American* article. Bryan reported that Clark "made me feel that he considered the 2 months with you as a 'matter of course' and not a 'matter of consequence.'"[91] Bryan had pressed Clark further, asking him, "I suppose you use Mr Akeley's method."[92] Clark's reply was evasive, indicating that the management thought Akeley's method "too slow & expensive."[93] Clark explained that he had modified the method by casting the manikin in plaster and by using "gas pipe for the supporting irons." Akeley later mused that his methods, "in the words of O. Henry, 'were damaged by improvements.'"[94]

Bryan emphasized to Akeley that the Clark incident "was *not* accidental"[95] and again urged his friend to publish the Akeley method. After "visiting the hole for the *new* Carnegie Museum, the new Washington Museum [the U.S. National Museum]—the proposed *new* Brooklyn museum," he was certain that if Akeley did not publish, "before the snow flies—they will have those ideas away from you—in some unknown *way* and will be using them in these museums—*Now Publish* them *at once*."[96] He further warned:

> And Akeley—that idea as you have shown it to me there—is in the *atmosphere*—The construction of any aquarium is along the same line!!— except for the backgrounds and I hold my breath for you for fear some one will tumble on to it either accidentally or through cross talk among architects, scientists or *some way*—and *thus* beat you out of the *"Akeley idea"* in museum construction—There is nothing like it in the world today—but I see it coming—[97]

Yet, for whatever reason, Akeley did not heed Bryan's warnings. Instead, he and Elliot planned another expedition to Africa. The FMNH administration agreed to finance a second trip, by which it hoped to secure elephants for the museum. When Elliot, at seventy, decided not to make the trip again, Akeley was made the expedition leader. He and Delia left Chicago for British East Africa (now Kenya) on August 13, 1905.[98]

While Akeley was out of the country, the museum world was shaken by the sudden death of Henry A. Ward.[99] Ironically, the man who had made his career from scientific and technological advancements was killed by the latest invention—an automobile—while visiting his family in Buffalo. Despite Ward's untimely death, his influence on America's museums was readily apparent. At the time of his death, in the summer of 1906, former taxidermists from Ward's shop occupied top-level positions at nearly all of the nation's major natural history museums and two of its largest zoos, as well as its largest aquarium. Ward's son, Henry L. Ward, was director of the Milwaukee Public Museum; Frederic A. Lucas was director of the Brooklyn Museum, where he employed William Critchley as his chief taxidermist and George K. Cherrie as curator of ornithology; at the New York Zoological Society, William T. Hornaday was director of the Bronx Zoo and Charles H. Townsend, director of the New York Aquarium; Arthur B. Baker was assistant superintendent of the National Zoo; at the U.S. National Museum, George B. Turner was taxidermist of mammals and Nelson R. Wood, taxidermist of birds; Harry C. Denslow was the bird taxidermist at the AMNH; and, of course, Webster was preparator at the CMNH, along with Remi Santens; and Akeley was chief taxidermist at the FMNH. Already these former colleagues had challenged one another to great heights, but soon their competitions and collaborations would encourage them to rethink fully the possibilities of museum display and what the very mission of natural history museums should be.

THE DEATH OF DECORATIVE TAXIDERMY

In the fall of 1906, CMNH director William J. Holland hired a new taxidermist to co-direct the preparations department with Webster. Remi Henri Santens was a recent Belgian immigrant who, in 1888, at the age of nineteen, began working in the taxidermy shop at Ward's Natural Science Establishment. He remained at Ward's for eighteen years, for the last nine of which he was co-foreman of the shop with his brother Joseph. In the estimation of Robert H. Rockwell, who apprenticed under the brothers at Ward's in 1905 and thereafter went on to a distinguished career at the U.S. National Museum, Brooklyn Museum, and AMNH, they "produced some of the best work in taxidermy done at Ward's." In fact, Rockwell admitted that their work was "so superior to mine that it didn't take a moment for me to realize that I had best scrap most of the mounting skills I was using and apply their technique."[100] Just six months after hiring Remi, Holland wrote that "Mr. Santens . . . possesses an excellent reputation as a skilful and conscientious workman. He has justified his reputation since with us by the work which he has done."[101]

Holland may have hired Santens to mollify W. E. Clyde Todd, custodian in charge of the collections of recent vertebrates, who had objected to

Webster's careless treatment of the mammal collection. In February 1900, only two years after Webster began working for the CMNH, Todd submitted a formal report to Holland, complaining about the condition of specimens entrusted to Webster and demanding that the director take control of the situation. Holland, however, had ignored Todd's complaints for six years, on account of harboring a rather open dislike for him. The tension between the two began after Holland refused to send Todd on various collecting expeditions or allow him time to publish his research. According to Kenneth C. Parks, "Todd always believed that whatever he accomplished had been done, not only without Dr. Holland's help, but in spite of his interferences."[102] But the problem did not go away, and Webster's mounts paled next to Akeley's. By hiring Santens, Holland could refocus Webster's attention on preparing bird specimens, while Santens mounted mammals.

It was not long before Santens brought further evidence that Webster had allowed the collections to deteriorate significantly. Two months after joining the museum, he submitted "a report of the condition of the skins in the Taxidermic Department," in which he stated that the skins had been "unmercifully treated" by Webster. "For nearly nine years skins have not been beamed nor dressed, large quantities of meat[,] grease and blood have been left on these skins. Just as animals were skinned, the skins were thrown into various baths solutions."[103] Santens was particularly appalled at the use of sulphuric acid on the skins. As a result, two elephants, three lions, a baboon, and a moose were no longer of any use in a scientific museum.[104] Other skins were forgotten "in salt and alum for so many years that they are all dried up and the salt and alum is crystallized on the hair and the original color is entirely ruined."[105] With Santens's claims to bolster his own, Todd once again attempted to force Holland into action. He submitted another report in January 1907:

> Recently the larger mammal skins have been overhauled and memoranda made as to their condition, with startling results. Of the skins preserved in bath 65 percent are practically ruined, 24 percent are doubtful, and only 11 percent are still apparently all right. The skins immersed in salt and alum bath have apparently fared better than those kept in a solution of which sulphuric acid was also a constituent. . . . [on] two elephant skins, for example, the epidermis is crumbling to pieces, and the specimens are entirely ruined.[106]

Todd wanted Holland to take control of the collection of mammals and appoint someone—though not himself—to oversee Webster's work. Under mounting pressure, Webster was forced to resign later that year. He accepted a position as superintendent of the Pittsburgh Newsboys Home, a nonprofit corporation for homeless newsboys. Joseph Santens, Remi's brother, was hired to replace Webster.[107]

Webster's forced resignation illustrates the degree to which museums were beginning to specialize. For years, Holland had relied on Webster—primarily a bird taxidermist—for an enormous variety of work: to mount taxidermy displays, prepare scientific specimens, and organize the collections. In reality, once the exhibits for the bird hall were completed, Webster was out of his depth. He had mounted some large mammals at Ward's, but the methods in which he had trained decades earlier had already been replaced and were now unsatisfactory for mounting large mammals. Remi and Joseph Santens, on the other hand, had trained at Ward's with Akeley, and thus had the ability to mount specimens by the new method. Versed in the latest preparation techniques, they were eager to design groups of mammals. Webster was no longer the best taxidermist for the museum.

AKELEY LOSES HIS CHAMPION

In 1907, upon returning to New York from Africa, Carl and Delia Akeley received word that President Roosevelt wished to meet with them. A few days later, they attended a luncheon at the White House, where a sportsman who had just arrived from Alaska was also in attendance. Although there was much discussion of Alaska and Africa, Akeley recalled that as they were leaving, Roosevelt turned to Delia and said, "When I finish this job I am going to Africa." The sportsman interjected, "How about Alaska?" Roosevelt replied, "Alaska can wait." The Akeleys' stories of collecting specimens in the Congo must have reawakened in Roosevelt memories of traveling with his father in Egypt; at age fourteen he had made a collection of the birds of the Nile Valley, which he later donated to the U.S. National Museum. Two years later, when Roosevelt retired from the presidency, he embarked on a year-long collecting expedition to British East Africa sponsored by the Smithsonian Institution.[108]

Akeley returned to Chicago on February 9 with more than seven thousand specimens, including four hundred skins of large mammals—seventeen tons in all, requiring 210 men to carry.[109] "The largest of their victims," reported the *Tribune*, "the elephants, are four in number, all but one of them having been secured in the neighborhood of Mount Kenya, a snow capped peak almost under the equator, where the hunters spent six weeks among the bamboo thickets."[110]

Fig. 4.14. Delia Akeley with African elephant skull and tusks, Chicago. (Image #46353, American Museum of Natural History Library)

Akeley's return to the Field Museum, however, was not the triumphant one he might have expected. While he was away in Africa, H. N. Higinbotham, president of the FMNH executive committee, had informed Elliot that his services were no longer needed at the museum. Elliot's position as curator of zoology had been given to Charles Barney Cory.

Cory was a dilettante naturalist from New England whose wealth allowed him to pursue an avid interest in bird collecting. He had a long history with the museum. When the Field Museum of Natural History was founded, members of the board who were friends of Cory asked him to donate his bird collection to the museum. Cory agreed, but asked in return to be made the honorary curator of ornithology. In 1906, when Cory lost his fortune to bad investments, the Field Museum's administration responded by giving him Elliot's paid position. The news stunned not only Elliot, but much of the staff. Many predicted that the transition would not be easy—that Akeley would be contentious when he learned that his friend had been pushed out and that Cory was now his supervisor.[111]

Before he left for Africa, Akeley had submitted a proposal for an "Exhibition of the Birds of Illinois." The proposal outlined an ambitious design:

> An exhibition of the birds of the state should contain as its main feature at least a pair, male and female, in breeding plumage, of every species found breeding within its limits, together with the nest and eggs of each, and a representative of its characteristic environment. Complemental to this main exhibit, there should be a collection of specimens and printed and photographic data available for persons desiring to know something of: (1) the life-history of each species; (2) its specific and individual peculiarities; (3) its seasonal distribution; (4) its economic relations; and (5) its psychological aspects.

Such an inclusive display, Akeley suggested, would require a "preliminary survey" of the state to determine which localities would be included in the final exhibit; at the same time, collectors could acquire representative accessories. Akeley also argued that the work of "the main or spectacular exhibit" should be the responsibility of the department of preparations, while the "subsidiary or educational exhibit" should be the responsibility of the scientific staff—in this case, the department of ornithology.[112]

Now, upon Akeley's return, he found that Elliot's replacement had rejected the proposed hall of birds. Akeley's suggestion to Skiff that the museum include in its Africa hall not two but five elephants was also rejected. Akeley was decidedly frustrated with the administration's lack of support for his ideas, its reluctance to expand the museum's public exhibits, and its growing disinterest in his experimental taxidermy. Just as he had at the Milwaukee Public Museum after Wheeler resigned, Akeley once again found himself without a director to champion his cause.

Although he did not know it at the time, Akeley's innovative concepts were beginning to make waves throughout the museum community. Museum directors across the United States were endeavoring to professionalize and had recently formed the American Association of Museums (AAM). Akeley soon discovered that there were more than a few museum directors who would stand staunchly in support of his modern approach to museum display.[113] In June 1907, Akeley attended the second annual meeting of the AAM in Pittsburgh, sponsored by the Carnegie Museum of Natural History. The association's first president, Hermon C. Bumpus, referred to Akeley as "one of the best taxidermists in America," and Henry L. Ward, before presenting a paper on the subject of exhibiting large animal groups, demurred that "Mr. Akeley, who had had as extensive experience in this matter as any man living," was better qualified than himself to speak on the subject.[114]

One year later, in early May, the AAM held its third annual meeting in Fullerton Hall at the Art Institute of Chicago.[115] Now that the nation's museum professionals had come to him, Akeley took the time to reiterate his method for mounting and exhibiting large mammals. Although his talk was not published in the *Proceedings of the AAM*, a brief mention was made of it: "Carl Akeley read a paper, illustrated by lantern slides, on modern methods in taxidermy."[116] Following Akeley's presentation, Henry L. Ward addressed the association:

> I do not know how this address of Mr. Akeley's has appealed to you. To me it seems to be epoch making. . . . I feel that tonight we have seen a breaking through of the old barriers of secretiveness that have not been a credit or an advantage to the individual taxidermist, to the museum that employed him, or to the art that he represented.[117]

Akeley had found many new champions beyond the walls of the Field Museum. With the assistance of Henry L. Ward, the newly formed association of museum professionals embraced his radical ideas, envisioning a new future for museum display.

But Akeley always operated more comfortably on the fringe, independently, never fully accepting a position within the profession. He saw himself as an idea man, and there was work to be done. By October, Skiff announced Akeley's resignation, citing his "desire to return to Africa, and a realization of the fact that a better field presented itself as an independent taxidermist."[118] Akeley would continue on contract with the museum to complete the pair of elephants for the entry hall, but if he ever intended to mount a large group of elephants, he would have to find support elsewhere.

Fig. 4.15. The armature for one of the pair of fighting bull elephants mounted by Carl E. Akeley for the Field Museum of Natural History in Chicago. (Image #410833, American Museum of Natural History Library)

He didn't have to wait long. Hermon Bumpus of the American Museum, a keen supporter of Akeley's work, had eagerly awaited the outcome of the Field Museum's elephant group.

On July 24, 1909, Akeley's fighting bull elephants were placed on exhibit in the main hall of the FMNH. According to the *Chicago Tribune*, Akeley "intended to surround the group with a reproduction of a jungle scene, but this would have involved an expense greater than the museum authorities had at their disposal."[119] The two elephants, in the end, disappointed Akeley—they were not the group he had hoped to mount—but others did not share his harsh opinion. Hornaday, for one, was especially vocal in his praise:

> Judged by the standards of artistic conception, this group is truly overwhelming. It is only by an effort that the imagination rises to its level, and yields to it the vast admiration that it deserves. It represents a gigantic conception and artistic effort successfully realized. It needs to be shown in a court at least a hundred feet square. It is a magnificent production, but, like the Sphinx among sculptures, it is not comparable with smaller creations of a pictorial nature. It is in a big and new class quite by itself.[120]

In the museum community, such enthusiasm for the fighting bulls created an instant demand for this new class of exhibits. In the spirit of competition, the AMNH was determined to have its own elephant exhibit, but unlike the Field Museum, it was willing to entertain Akeley's idea of designing a large family group.

Without hesitation, the American Museum hired Akeley to mount the world's first large group of elephants and promised to fund his third expedition to Africa. According to *Science*, the expedition was expected to take two years, and in addition to collecting "a group of elephants to be mounted amid a reproduction of their natural habitat in the American Museum," Akeley was also tasked with "making a very complete photographic record of the people, fauna and flora" to use for research and educational purposes.[121] Although he successfully made stills and motion pictures of animals in their natural habitats, such as weaver finches, ostriches, elephants, baboons, and for the first time ever, gorillas, Akeley found his Urban motion picture camera to be heavy, awkward, and cumbersome to use in the field. To solve this problem, Akeley would later design and patent the more portable 35 mm Akeley Motion Picture Camera, which quickly became the camera of choice for field naturalists.[122]

Fig. 4.16. Carl E. Akeley's pair of fighting bull elephants mounted for the Field Museum. (Courtesy, Field Museum. Image #CSZ29279)

Akeley left for British East Africa on August 17, 1909. Theodore Roosevelt, making good on his promise to Carl and Delia two years earlier, had arrived ahead of them in the port city of Mombasa in April. Although Akeley had declined to join Roosevelt's Smithsonian Institution–sponsored expedition—as he was already committed to the AMNH—he did arrange for the president to use his own trusted outfitters, Newland & Tarlton. They had successfully handled the logistics for Akeley's 1905 expedition, and he intended to use them again for the AMNH expedition. Akeley also suggested that Newland & Tarlton make arrangements for the two expeditions to rendezvous at some point so that he and Roosevelt could hunt elephants together, as Akeley had convinced Roosevelt to collect two elephant cows and a calf for the AMNH.[123] When Akeley and Delia arrived in Nairobi, a letter from Roosevelt was waiting for them: the expeditions would rendezvous on the grasslands of the Uasin Gishu Plateau. As the

Akeley expedition made its way across British East Africa, Akeley received periodic updates about Roosevelt's whereabouts from African runners who were familiar with the area and hired to communicate the location of each group. Once Akeley's expedition reached the plateau, a runner set out to find Roosevelt, who, it turns out, was not far away.

Akeley and Roosevelt set up a camp from which to hunt elephants. The next morning, Akeley, Tarlton, Roosevelt, his son Kermit, and a group of African gun bearers and trackers set out in search of elephants. By noon, the group had walked nearly ten miles before they spotted a herd of elephants gathered in the long grass near a grove of small, sparsely leaved mimosa trees. Roosevelt was allowed to take the first shot—striking one and then a second cow—and then Kermit, Akeley, and Tarlton followed suit. By the time the dust from the charging herd settled, Roosevelt had shot three cows, and Kermit, with his Winchester rifle, had killed a bull calf. After Roosevelt and Kermit posed for photos, Akeley and Tarlton immediately set to skinning the elephants. By evening, the grisly task was complete. In the morning, when Akeley's assistant, James L. Clark, arrived (after a night lost in the bush) with the necessary supply of salt, the two men, together with a few African assistants, began the hard work of cleaning and preserving the hides. It wasn't long before the men became aware of a fast-moving brush fire that threatened their camp. Akeley continued to work on the hides through the smoke and heat while the rest of the men successfully fought back the flames. But the delays, coupled with the scorching sun, caused the skins to begin to rot. Akeley had to move fast. They splashed saltwater over the hides, which allowed them the necessary amount of time to remove the fats so that the skin could begin to absorb the salt. To ensure that the hides would remain secure from rot, Akeley took the additional precaution of packing them in cotton cloth impregnated with beeswax before crating them for shipment to the AMNH. The following day, Roosevelt parted company with the Akeleys and headed toward Lake Sergoi. The Akeley expedition would eventually find a bull elephant in Uganda to complete the AMNH elephant group.[124]

When Akeley returned from British East Africa in 1911, he arrived in New York with the hides necessary to mount the first complete elephant group, but his experience observing live African animals, particularly elephants, had left him "dreaming of a great African Hall which would combine all the advances that had been made in taxidermy and the arts of museum exhibition and at the same time would make a permanent record of the fast-disappearing wild life of that most interesting animal kingdom, Africa."[125]

Akeley's development of a new concept of museum display that successfully merged scientific and decorative taxidermy coincided not only with the rise of public exhibition space in natural history museums, but also with a growing wildlife conservation movement in the United States. This confluence promised museum displays that combined all the advanced techniques of taxidermy with the latest scientific information—presented in an accessible language meant to inform, and in some instances influence, museum visitors. Furthermore, with the newly defined central place of habitat groups in public exhibition, museum administrators were open to rethinking the nature of their collections. They would eventually divide their specimens between public displays and scientific collections. This division further reshaped the museum institution and led administrators to reconsider the educational value of their exhibits—adapting not only the taxidermy, but also the arrangement of halls, exhibit labels, and accompanying materials, for the benefit of their growing public audience. Ultimately, it was through the adoption of Akeley's museum methods and ideas as the new vision guiding the creation of exhibition spaces that taxidermy came to shape and define the public aspect of American natural history museums.

CHAPTER FIVE

"The Duty to Conserve": Museums and the Fight to Save Endangered Marine Mammals

We are weary of witnessing the greed, selfishness and cruelty of "civilized" man toward the wild creatures of the earth. We are sick of tales of slaughter and pictures of carnage. It is time for sweeping Reformation; and that is precisely what we now demand.
—William T. Hornaday[1]

On the afternoon of February 20, 1904, a large crowd packed into the lecture room at the U.S. National Museum to hear a visiting lecturer, Charles H. Townsend, director of the New York Aquarium, describe the mysteries of the ocean floor. While still the naturalist aboard the U.S. Fish Commission's ship *Albatross*, Townsend had participated in expeditions taking soundings of the ocean floor from San Francisco to Honolulu, measuring distances as much as six miles below the surface of the waves. "The sunlight penetrates the ocean to a depth of 200 fathoms. Below that, to the bed of the ocean, is darkness," Townsend told the audience. "At the bottom, however, life is abundant. There is no sunlight. All is darkness and intense cold, but the fishes and other animal life are provided with phosphorescent light of their own, by which they are able to see. We believe that the sea bed is illuminated by a glow from the animal life inhabiting these depths."[2]

The public knew relatively little about what lay beneath the surface of the world's oceans, and they listened in awe as Townsend declared that a new, "heretofore unknown field for the naturalist has been opened."[3] He explained that no deep-sea creatures could survive the journey topside, but that by studying the anatomy of such bottom-dwellers and making careful reconstructions, we could begin to imagine what their hidden world must look like. To offer a glimpse of that habitat, Townsend showed slides of several exhibits that the Smithsonian was constructing for the upcoming

World's Fair in St. Louis. In surveying the lecture hall, a reporter for the *Washington Post* noted a conspicuous number of rapt children in the audience. "They were unusually well-behaved," he wrote,

> their attention being riveted to the lantern slide pictures of wonderful deep-sea monsters. Yet these children doubtless little suspected that in the workshop of the ichthyological taxidermist, situated within half a block of the museum, were many of the originals of the strange fishes, shown together with enlarged models, either completed or unfinished, which are to form part and parcel of the museum's exhibit at St. Louis.[4]

Even those who may have known their proximity to the museum's workshops could not have guessed the ambition of the Smithsonian's undertaking.

Upon being awarded the honor of holding the World's Fair for 1904, the United States had chosen St. Louis, gateway to the American West, as the host city, and had officially dubbed the celebration the Louisiana Purchase Exposition to honor the centennial (one year late) of President Thomas Jefferson's acquisition of the western territories in 1803. The whole country expected a show that would outdo the extravagance of the 1893 Chicago World's Fair, and Frederic Lucas, now head of exhibits at the National Museum and in charge of preparing the government's exhibits, intended to give it to them. In Chicago, Lucas had unveiled the first accurate depiction of an underwater seascape by modeling a group of octopi in clay, then casting them in a mixture of glue and gelatin.[5] Now, a decade later, he planned to give fairgoers a view into an underwater realm more extensive and more realistic than any they had ever seen before—crowned by the world's first full cast of a whale.

Lucas hoped that the exhibits would also serve to educate the masses about the need for species conservation even in the oceans. Since serving with Townsend on the Fur Seal Commission of 1896–1897, Lucas had become increasingly aware of the rapid depletion of marine mammals in particular. However, it was clear to Lucas that it would be challenging to raise awareness about the plight of species less iconic than the American bison. Many aquatic animals lived in remote locations—on inaccessible islands, in polar regions, or at home under the waves—which made them more difficult to study and preserve; as such, they were not often seen in museum exhibits. Without immediate action, Lucas feared that the fate of the great auk was soon to befall a shocking number of marine mammals—the sea otter, the walrus, the northern fur seal, and various species of whales. Townsend joined Lucas in his conservation work—and soon expanded it to include issues of water pollution. However, when both men concluded that the best way to revive the dwindling herds of fur seals was through selective management, they encountered fierce opposition from their old ally William T. Hornaday. The battle that followed threatened to divide the conservation movement and continues to shape arguments between conservationists and advocates of wildlife management to this day.

THE MAKING OF A WHALE

Frederic Lucas arrived at the docks in the capital city of St. John's, Newfoundland, in the middle of May 1903. After disembarking, he and his companions quickly made their way down the main thoroughfare of Water Street; they were in need of an odd assortment of supplies, including twenty barrels of plaster of Paris. Lucas had been detailed by Frederick W. True, curator of mammals at the U.S. National Museum, to lead an expedition to collect a specimen of the blue whale—then referred to commonly as the "sulphur-bottom whale." Lucas was accompanied by William Palmer, who had replaced Hornaday as chief taxidermist, and J. W. Scollick, osteological preparator and second-generation graduate of Ward's Natural Science Establishment. Lucas had not had many opportunities in the field, and as he later recalled, "The worst part of this work was the fear of failure, the worry lest we did not get a good big whale, and there were some anxious days when no sulphur-bottoms were taken or even seen."[6] For nearly four weeks, they waited at the Cabot Steam Whaling Company's principal station, Balena, on Hermitage Bay, for the perfect specimen.

Whaling stations had been established near whale breeding grounds along the shoreline of Newfoundland, growing from one in 1897 to five in 1903. Little was known about whale biology at this time, but Lucas understood that the species prized by the whaling industry were experiencing devastating declines in their populations. Right and bowhead whales had been hunted to near extinction, and now that these species were harder to find, the whaling industry had turned its sights on the blue whale. In early times, the fast-swimming blues had been impossible to hunt effectively, but technological advances had caught up to them—with steam-powered ships now receiving direction from lookouts at ground stations and hunting with exploding harpoons. Lucas had learned from Hornaday that if he wanted to rally the American public to save whales from extinction, he would first have to obtain a specimen and mount an educational exhibit. The blue whale was the obvious choice for his representative species, as its size alone would captivate museum visitors, and it would also complete the museum's collection of cetaceans.[7]

Fig. 5.1. Joseph Palmer's half-cast and skeleton of a humpback whale, shortly after it was installed in the U.S. National Museum in 1885. (Smithsonian Institution Archives. Image #2002-12204)

However, Lucas had another, more personal reason for obtaining a complete cast of a blue whale. Five years earlier, he had read with dismay an editorial notice in the July 1898 issue of *Natural Science*, an upstart British scientific monthly, describing the British Museum's new Cetacean Gallery. The author claimed that no other museum had "solved the difficulty of exhibiting the outward form of the various kinds of whales which baffle the taxidermist's art," until Sir William Henry Flower had "at last . . . solved the problem in a most satisfactory manner."[8] Although the British Museum was well known for its airs of superiority, Lucas was incensed by the claim, as he knew that the Smithsonian's longtime modeler and taxidermist, Joseph Palmer, and his son, William Palmer, together with Secretary Spencer F. Baird, had discovered this exact solution sixteen years earlier while mounting the skeleton and cast of a thirty-three-foot humpback whale. The innovative cast revealed on its left side a humpback whale in an "attitude of swimming through the water," while the right side exposed the animal's complete articulated skeleton.[9] The half cast had been on exhibit in the south main hall of the National Museum since 1882.

To challenge the misstatement, Lucas urged Frederick W. True, a scientific authority on whales, to pen a response. *Science*, eager to establish preeminence for the U.S. National Museum over the British Museum, rushed to print the editorial, which appeared in the July 22 issue. True explained that the method of "exhibiting *papier maché* casts of one-half of the exterior of the various cetaceans, colored as in life, and placing the skeletons in the concavities of the casts" was an idea that had originated at the U.S. National Museum and had "been in use . . . for more than fifteen years." Although True stated that he was certain that Flower would "disclaim originality for this excellent mode of exhibiting cetaceans," he clearly meant to embarrass him, quoting a passage from the Smithsonian's annual report for 1882 that described the new mode of displaying whales, and enumerating the several mounted specimens of smaller whales that the National Museum had exhibited since 1874, and indeed, at the London Fisheries Exhibition in 1883, after which several of the mounts were given to the British Museum by Flower's American counterpart, George Brown Goode.[10]

Lucas and True were apparently not satisfied to end this transatlantic rivalry with an editorial: the Newfoundland expedition was to be the coup de grâce. Lucas had created an entirely new method for obtaining a scientifically accurate and complete cast of a whale. The old method was to take casts of the animals "lying high and dry on the beach," without accurate measurements. Given the enormous size of many whales, it was impossible to turn the specimen over to acquire a true cast of its alternate side; as a result, a cast was taken of only one side, and the other was "built up from one cast, and filled out or altered wherever discrepancies happened to exist in making the two sides from one." Lucas believed the solution was to cast the dead animal while floating it in water—but weeks passed without a suitable specimen on which to test this new technique.

Finally, on July 12—nearly two months after Lucas arrived in Newfoundland—the Balena station received word that one of its steamers had hauled in a blue whale, measuring seventy-eight feet in length and weighing seventy tons. Thrilled by the catch of a fine specimen, Lucas instructed the captain to tow the body "into shoal water [about ten or twelve feet deep] just as the ebb tide set in." On shore, Lucas, Palmer, and Scollick directed four of the company's employees to begin mixing plaster of Paris and excelsior with water in the large wooden vats they had built for the occasion. In

Fig. 5.2. Frederic A. Lucas (*left*, atop whale), J. W. Scollick (*middle*), and William Palmer (*right*) plastering the head of the blue whale, 1903. (Smithsonian Institution Archives. Image #SIA-2012-6537)

the water were two poles between which the animal would rest. Attached to the poles was a staging area from which to cast the higher portions of the body and a system of ropes that would be used to turn the animal. Once the whale was in position, "tail toward the beach and the head seaward," resting on its left side, the three waded out into the frigid water and "commenced work with a vim." For the next ten hours, the station men carried buckets of plaster out to the museum workers, who poured it over the whale's body, as "the animal lay on its left side. As the tide fell, they worked down toward the median line [of the stomach] on each side," taking molds in sections, until the work on the body was complete. Whale flesh decomposes rapidly, so the exhausted group had to continue working until the entire cast was complete. They left the head, which decomposes more slowly than the rest of the body, for last. At that time, "the whale was hauled out on land and decapitated. . . . As soon as it was severed from the trunk we took a complete cast of the member, jaws and all, both inside as well as out," and the flukes were molded separately. Lucas recalled that "it was the hardest work

I ever performed." For the next several days, the station men helped strip fat from whale's skeleton, while Palmer and Scollick dismembered it; Lucas was determined that "every part of the whale's frame, even down to the smallest and most minute bones," would be collected and treated with care.

The expedition returned to Washington on July 22, with the skeleton and molds in several large crates. Lucas oversaw the modeling of the specimen in an enormous shed, built especially for the purpose. As he hoped to have it completed in time for the St. Louis exposition the following year, he wasted little time in staging the work. On August 16, the *Washington Post* reported:

> Those who are anxious to settle the problem whether Jonah was actually swallowed by a whale would do well to pay a visit to the rear of the Smithsonian Institution, where for some days past a most remarkable and peculiar diagram has remained staked out on the lawn . . . that has

MAKING PLASTER MOLD OF BODY OF SULPHUR-BOTTOM WHALE

Fig. 5.3. Frederic A. Lucas (atop whale) and J. W. Scollick (*third from left*, leaning against whale) preparing the cast of the blue whale's body, 1903. (Smithsonian Institution Archives. Image #SIA-2012-6539)

SKULL OF SULPHUR-BOTTOM WHALE CRATED FOR SHIPMENT TO WASHINGTON

Fig. 5.4. Frederic A. Lucas (*second from right*) with skull of blue whale crated for shipment to the U.S. National Museum from Newfoundland. (Smithsonian Institution
Archives. Image #SIA-2012-6536)

greatly puzzled those who have occasion to cross the Mall. . . . Prof. Lucas . . . and Mr. Palmer, the chief preparator, were bossing the work, and
the diagram in question was the lateral outline and proportions of a big
whale. . . . The diagram, as laid out, gives one a very correct idea of the
whale, and those anxious to ascertain the truth or falsity of the Jonah
story are at liberty to measure their length on the well-kept lawn within
the area marked off for the whale's stomach.[11]

News of the National Museum's "cetacean monster" captured the American public's imagination as accounts appeared in numerous newspapers. So
intense was the curiosity that in November, Lucas presented a talk titled
"The Making of a Whale," illustrated by Palmer's slides, to the Biological
Society of Washington, in which he explained the entire process, from the
making of the molds to the planned casting.

In the end, it took eight months to complete the enormous manikin.
The exterior was made of papier-mâché, using old paper money pulp from

Fig. 5.5. Workers in the South Yard behind the Smithsonian Building making the papier mâché model of the body of the blue whale for the Louisiana Purchase Exposition, 1903. (Smithsonian Institution Archives. Image #82-3371)

the U.S. Treasury, and painted by Palmer. In early March, the whole was disassembled into sections and crated to be shipped by rail to St. Louis. The head filled one crate, the flippers another, the sides two more, and the tail a fifth, collectively covering a flatcar, over which a protective housing was erected. After so much anticipation, the *Washington Post* marveled at the lack of fanfare for the whale's departure. "The switchmen, freight handlers, draymen, and shipping clerks little imagined the nature of the contents of the immense crates," the writer reported, "and on the whole the whale managed to slip out of town with very little excitement."[12]

At the St. Louis World's Fair, the Smithsonian coordinated all of the exhibits for the U.S. government. Its building was considered one of the most impressive at the fair and featured a full-sized model of the U.S. battleship *Missouri*; a relief map of the projected Panama Canal; a live fish exhibit with fifty aquaria containing freshwater and saltwater species; and "the largest bird cage ever built," where visitors encountered hundreds of bird species from around the globe. The immense blue whale cast, which hung from the rafters, was described "as the most striking object . . . showing the natural appearance of this greatest of all living creatures."[13]

When the blue whale cast returned from St. Louis in 1905, it was suspended from the roof trusses of the Smithsonian's South Hall in the Arts and Industries Building. After the new U.S. National Museum building opened in 1910, it was moved across the National Mall, mounted on a pedestal, and placed at the center of the Hall of Marine Life. For fifty years, the seventy-eight-foot cast of the blue whale enchanted visitors to the museum.

After seeing the mount of the blue whale completed and safely shipped to St. Louis, Lucas left the National Museum to assume the position of

Fig. 5.6. Frederic A. Lucas's full cast of the blue whale and its skeleton, as installed at the Louisiana Purchase Exposition in St. Louis, 1904. (Smithsonian Institution Archives. Image #NHB-16424)

FULL-SIZE MODEL OF A SULPHUR-BOTTOM WHALE, 78 FEET LONG

Fig. 5.7. Director Remington Kellogg and Leonhard Stejneger view Frederic A. Lucas's full-sized model of a blue whale, seventy-eight feet long, on exhibit in the Museum of Natural History, ca. 1930s. (Smithsonian Institution Archives. Image #SIA-2012-6538)

curator-in-chief of the Brooklyn Museum, a division of the Brooklyn Institute of Arts and Sciences. But he had not forgotten what he saw on Newfoundland—nor the long wait he had to endure as his crew hoped for the arrival of a blue whale carcass. In 1903, R. T. McGrath, in his *Report of the Newfoundland Department of Fisheries*, had cautioned against granting licenses for further whaling in the area. "It will result in the complete depletion of this industry within a short time,"[14] he wrote, but his advice was ignored, and Lucas feared that the warning had come too late. In 1906, the number of whales caught dropped precipitously—to exactly half the number caught off the coast of Newfoundland just three years before, despite the fact that there were now five times as many stations.[15]

The conclusion was inescapable: the blue whale was rapidly headed for extermination. "Man is recklessly spending the capital Nature has been centuries in acquiring," Lucas wrote in his disquisition on the need to protect whale species, "and the time will come when his drafts will no longer be honored." Worse still, the Newfoundland industry was only one part of a larger picture; new, more advanced whaling stations were rapidly being established all over the world. If the problem was to be solved, it would take more than public outcry, more even than state or national laws. "Whales can be protected and protected very easily," Lucas concluded, "but it can only be done by international agreement."[16]

To achieve this difficult goal, Lucas enlisted the assistance of his old friend Charles H. Townsend, the first director of the New York Aquarium under the auspices of the New York Zoological Society. Townsend successfully convinced the NYZS to adopt a resolution calling for the protection of whales by international agreement.[17] He also published Lucas's "The Passing of the Whale" as a special supplement to the *Zoological Society Bulletin*, which Townsend distributed to members with his endorsement, calling Lucas's report "a truthful statement by one of the best-informed students of the subject." The supplement was also distributed to popular magazines for reprinting and sent to "legislative bodies in many parts of the world." Townsend, as director of the aquarium and representative of the NYZS, which he identified as "a scientific association devoted to the preservation of wild animals," urged "the careful consideration of it by every legislator into whose hands it may come."[18]

TOWNSEND AND THE AQUARIUM

In 1902, the Municipality of New York invited the New York Zoological Society to assume control of the New York Aquarium, then located in Castle Garden at Battery Park. "After some deliberation," reported society president Henry Fairfield Osborn, "the invitation was accepted, the necessary legislation at Albany was secured, and a contract was made."[19] Osborn offered the new position of aquarium director to Charles Townsend. Townsend was eminently qualified, having received instruction in oceanography from Alexander Agassiz while serving under him aboard the U.S. Fish Commission steamer *Albatross* in 1891, and having pursued his own research as naturalist aboard the same ship for the next decade. As he later recalled, "The long voyages with Agassiz, always illuminated with his enthusiastic talk in the ship's laboratory, amounted to a course in oceanography . . . and eventually anchored me at the Aquarium."[20] In November, Townsend resigned from the Fish Commission to become director of the aquarium, and almost immediately was sent on a tour of European aquaria with "a view to studying the best foreign methods."[21]

In England, he visited the facilities at Plymouth and Brighton, then continued on to Paris, Berlin, and Naples. Of these, Townsend was especially impressed by the Naples Aquarium, which he considered "far and away the finest, exhibiting semi-tropical fish taken out of the Mediterranean and the fish in a novel and artistic manner."[22] Upon his return to New York, he undertook a complete redesign of the aquarium's tanks, patterned after what he had seen in Naples. Under the new design, the tanks were to be

> backed up with natural rock. The beds of fresh-water streams will be duplicated as far as possible for the fish coming from those streams, and salt grottos will be built for the deep-sea fish. Seaweed and other marine plants will be cultivated in the tanks, giving the fish as nearly a natural surrounding as possible.[23]

To ensure absolute accuracy, these naturalistic settings were fashioned to represent actual locations shown in underwater photographs taken by Townsend and, whenever possible, furnished with native vegetation, rocks, and sand.[24] For the tank of Bermuda fishes, for example, coral was brought in from the reefs off the coast of Bermuda.[25]

Townsend also ordered the bare plaster interior of Castle Garden painted, large sections of the roof removed and replaced with skylights, and a new laboratory facility constructed for research and classroom instruction. But his concerns were more than cosmetic. Townsend believed that many of the saltwater fish that died each year were being killed by the harbor water that was pumped from wells under the aquarium directly into holding tanks. In the winter, when the Hudson swelled and backed up with snow and ice, freshwater in New York Harbor overwhelmed saltwater. To solve the problem, Townsend proposed that a 100,000-gallon tank be erected outside the aquarium at the west end of Battery Park. Basing his design on the reservoirs at the Naples Aquarium, he corresponded with its director, Anton Dohrn, regarding the desirable salinity levels for the water.[26] He even arranged to have the tank periodically filled by incoming ships that had taken on seawater as ballast.

To Townsend's distress, despite this mixing of water from the open ocean with filtered harbor water, the mortality rate at the aquarium continued to increase each year. He asked the Metropolitan Sewage Commission to investigate the harbor water. The results were shocking. As the *New York Times* reported in 1908, "A coating of sewage and factory waste several feet thick has formed over the bottom of New York Harbor and may become a menace to the health of the city. It has already destroyed most of the forms of marine life which assist in the disposal of organic matter at the harbor bottom." Worse still, due to increasing pollution, the layer was gradually growing.[27] At a meeting of the Angler's Club of New York, Townsend urged the group to "take a determined stand" on factory pollution of freshwater streams. "All the deleterious matter now liberated in our angling waters is destroying fish life," he told them, "and stream pollution, both in upper and lower waters, is a most serious source of injury to our fisheries."[28] In the meantime, Townsend implemented a plan to import hundreds of thousands of gallons of seawater from Bermuda and use a completely closed system for the saltwater fish.[29] Eight years later, although Townsend had continued his campaign to stop the flow of sewage into New York Harbor, the water quality still had not improved, and installation of the aquarium filtration system had not been completed. As a result, by 1916, the entire school of five bottle-nosed dolphins housed in the large central pool of the building had succumbed to waterborne diseases. Townsend deeply regretted the loss, as the males and females were often observed mating. He wrote, "The loss of the females was especially disappointing as the prospects for breeding in captivity were promising."[30]

Townsend came to realize that the captive breeding of marine mammals might not be an attainable goal. Despite his efforts to overcome the many issues associated with providing for marine mammals, including several species of seals, manatees, and bottle-nosed dolphins, the animals either died en route to the aquarium or lived for only a short time in captivity—on average, one year. Under such stressful conditions, it was impossible to encourage the animals to breed. However, Townsend continued to accept marine mammals when they became available. In 1909, the aquarium received an adult male and three yearling West Indian seals, a species that until recently had been thought extinct. Townsend described their fragile state upon arrival: "One of the latter was in a weak condition and died the day after arrival. The others are apparently doing well. . . . They are probably the only specimens of this nearly extinct species now living in captivity."[31] All died within months, but for Townsend, it was important to exhibit rare species of animals to educate museum visitors and encourage conservation, even if they might survive at the aquarium for only a short time.

LUCAS AT THE BROOKLYN
INSTITUTE OF ARTS AND SCIENCES

When Frederic Lucas arrived at the Brooklyn Institute of Arts and Sciences on June 1, 1904, he found the Department of Natural History's public exhibits in an impossible state of disarray. As he recalled later, the curators of geography, botany, mineralogy, and zoology had been vying for space and prominence in the exhibit halls until the museum had come to look "something like an ancient village site, in strata of disorder, the exhibits of the most recently active department being at the front and the others pushed to the back." As a result, he said, it was more than a year before he invited any friends to visit the museum, when "we had made appreciable progress and chaos was giving way to mere confusion."[32]

Ideally, Lucas sought to install habitat groups with elaborate surroundings and painted backdrops like Akeley's "Four Seasons" at the Field Museum. Such groups, however, required "considerable sums of money and a large efficient corps of workers," neither of which were at his disposal. His resources were scant, and he considered the majority of his staff "sometimes to be inefficient, and more often unsympathetic or interested only in their own work."[33] Fortunately, among the qualified staff Lucas had inherited were George K. Cherrie, an experienced field collector who had trained at Ward's, and J. William Critchley, a skilled if uninventive taxidermist who had trained at Ward's under Akeley. Almost immediately, Lucas dispatched Cherrie to South America to complete a series of New World birds; meanwhile, he set Critchley to mounting a group of northern fur seals that he had obtained "with the aid of the North American Commercial Company" through the Department of Commerce and Labor.[34]

The fur seal group, as Lucas conceived of it, was large and ambitious. Composed of thirteen individual specimens, the scene included an old male, two nearly grown "half bulls," three young males, two full-grown females, and five pups—"the whole," Lucas wrote at the time, "giving an extremely good idea of these interesting animals."[35] Critchley used Townsend's field photographs of fur seal rookeries in the Bering Sea as a guide to mount the specimens in a naturalistic pose. Turn-of-the-century technological advances in photography—increased portability of cameras, which allowed photographs to be taken in the field, and higher shutter speeds, which made it possible to photograph subjects in motion—gave taxidermists a better idea of an animal's behavior and corresponding natural attitudes. As such, for taxidermists like Critchley who had not observed the animals in the wild, photographs were an invaluable aid for mounting specimens in naturalistic poses.

Once they were completed, Lucas directed Critchley to arrange the fur seals on a simple base of artificial rocks in a style similar to Julius Stoerzer's group of seals mounted for the U.S. National Museum's exhibit at the Centennial Exhibition in Philadelphia nearly three decades earlier. However, whereas Stoerzer had arranged his seals as statues, without any apparent interaction among them, Lucas designed his group to narrate various aspects of behavior he had observed in the field. He described the scene:

> The most prominent figure in the seal group is an old bull, or fully-grown male, who is represented as being on the rookery ground threatening the younger "half bulls" with immediate death if they venture in his vicinity. One half bull, a four-year-old, in response to this admonition, is beating a retreat, while his five-year-old companion, encouraged by his position, tells the bull he is not afraid. . . . Coming from the sea, where she has been sleeping and feeding for a week or ten days, is a mother seal, or cow, calling to her young "pup," who runs bleating to her.[36]

In order to reduce expenses, instead of commissioning a painting to depict the background, Lucas displayed a composite of Townsend's photographs, "seven feet long showing a rookery half a mile long with two thousand seals upon it,"[37] with the mounted exhibit. Photographs, as Lucas found, not only gave the museum visitor "a far better idea of the facts, including the social environment and behavior of a particular species, but they were also much less expensive to produce" than a painted background,[38] In November 1905, Brooklyn's group of fur seals was unveiled on the second floor of the central section of the museum, and *Science* immediately declared it "the finest of its kind in any museum."[39]

Even with this stellar achievement, Lucas wasn't convinced that large habitat groups were the educational design solution for all natural history museums, as they required significant financial resources, large amounts of space, and talented taxidermists, designers, and artists. In June 1907, at the second annual meeting of the American Association of Museums (AAM) in Pittsburgh, as Henry L. Ward of the Milwaukee Public Museum presented a paper titled "The Exhibition of Large Groups," Lucas challenged the assembled museum builders to consider the question, "Will you endeavor to show the animals, or the conditions under which they occur?"[40] Lucas explained that he had long considered the "group question" and had come to the conclusion that "backgrounds should be entirely subordinated to the animals." A lively debate ensued.

Ward answered that he believed it was important for exhibits to show "the relation of the animal to its environment" and that this could best be accomplished with "pictorial backgrounds." He used sea lions as an example: "The closeness with which these animals are associated in a group would suggest their gregarious habits. . . . But if there be a background showing a herd painted upon it, the observer is impressed by the fact of their gregariousness in a way that four or five specimens put into a group cannot possibly cause him to be."[41]

Frederic Webster interjected that "with background or without," photographs might be useful as "supplementary instruction." Webster had just completed a pair of fur seals for the Carnegie Museum, donated in 1897 by Lucas, and he suggested a seal rookery as an apt example. If a few specimens were mounted to create a group, the social environment of the rookery could be shown in a photograph of the actual breeding grounds.[42] Lucas volunteered that the Brooklyn Museum's fur seal group had followed exactly this principle, with great success.

At last, Lucas had an exhibit at the Brooklyn Museum that could serve as a model, instead of an embarrassment. But, more than merely improving the quality of the museum's exhibits, Lucas hoped the fur seal group and its striking array of photographs would provide an opportunity to implement his ideas about museum labels. He liked to tell the story of a carpenter who was repairing a case at the Brooklyn Museum that held a common house cat; seeing that the label described the cat as a "member of the family Felidae, a group of carnivorous digitigrade mammals," the carpenter asked the curator overseeing his work to explain. The curator clarified that "the cat ate meat and walked on its toes." Lucas saw this story as representative of many lost opportunities within museums: if the public couldn't understand the labels, then they couldn't be informed—or persuaded. Thus, it became Lucas's principle never to have "'digitigrade mammals,' in place of plain cats."[43]

The educational message that Lucas hoped to get across with the fur seal group was an environmental one. He supplemented the mounted specimens with text describing pelagic sealing—the practice of shooting seals in the ocean—and emphasizing its destructiveness to seal populations: during breeding season, it resulted primarily in the death of nursing females, which hunt for food far from the rookeries while the males remain on land to protect the pups. The educational materials included

> a skin so dressed as to show the various stages in preparation; this, which hangs near by, is an object of much interest to visitors. The blank faces of the rockwork have been utilized for labels, and these comprise beside the general label, classes of seals, sealers terms, and pelagic sealing. One side is taken up by a fine panoramic view of Polovina Rookery with its

breeding seals, idle bulls and bachelors and on the end is a map showing the migrations of seals on both sides of the Pacific.[44]

The labels pointed out scientific inaccuracies in the mounted group, noting that "these seals would not be found so close to one another," but, Lucas lamented, "the limits of space available for the group has, as is often the case, caused truth and consistency to be sacrificed to convenience." The labels also noted that millions of dollars had been spent in an effort to end pelagic sealing. "The pity of it is that no animal can be so readily cared for as the fur seal, and the abolition of killing at sea would mean an assured supply of seal skins for all time." The text, however, also decried land killing, noting that the three bachelors grouped in one corner of the exhibit were "just below killable size, though in these days in danger of losing their hides to furnish gentle woman with an unnecessary cloak."[45]

Taken as a whole, the exhibit was intended to interest the public in the endangered northern fur seal, but it was also meant to persuade. Lucas and Townsend had been urging an international ban on pelagic sealing for close to a decade, but the plight of the seal had failed to capture the public's attention. Lucas's exhibit made the visual argument that pelagic sealing had forced the great herd of northern fur seals that once numbered in the millions to the brink of extinction. Within a few years, the general interest would grow—and the pressure of public scrutiny, together with the threat of economic impact, would finally force lawmakers to negotiate an international treaty to end pelagic sealing in the North Pacific. Lucas and Townsend would lead this effort, but they would face an unexpected opponent in their old friend William T. Hornaday.

FORMATION OF THE FUR SEAL ADVISORY BOARD

In January 1909, Townsend and Lucas were once again called upon to serve as scientific experts in the matter of the fur seal. Again serving under ichthyologist David Starr Jordan, they were appointed to the U.S. Department of Commerce and Labor's Fur Seal Advisory Board (FSAB). Consisting mostly of members of past commissions, including C. Hart Merriam, chief of the U.S. Department of Agriculture's Division of Biological Survey, and Leonhard Stejneger, curator of the Department of Reptiles at the U.S. National Museum, this loose-knit group of scientists had been working together studying the northern fur seal on and off for the last twenty

years. Nominally, the goal of this board was to consider the management
and preservation of the Pribilof herd and the advisability of continuing the
North American Commercial Company's hunting lease after its expiration
in April 1910; however, political pressure was now greater than ever for the
scientists to find a permanent resolution to the controversy, as public concern mounted over the impending extinction of the fur seal. "Many of the
problems with which this board will have to deal are of great importance,"
the *Washington Post* reported, "and their proper handling is essential to the
rehabilitation and preservation of the fur-seal herd."[46] But more than simply
preserving this species, the board realized that if scientists could resolve an
international dispute over economically valuable natural resources, they
would establish a precedent for the future involvement of the scientific
community in shaping public policy—an outcome that might save many
threatened species.

On June 29, the members of the FSAB departed Seattle aboard the steamship *Victoria*, bound for Nome, Alaska. Arriving ten days later, the crew
was transferred via the cutter *Rush* to St. Paul Island. For all of July and into
August, the scientists took stock of the breeding harems at the seventeen
fur seal rookeries on St. Paul and five more on St. George Island. Compared
with what these same men had seen only twelve years earlier, the herd was
now "a skeleton or outline, the substance having gone."[47]

Decades of pressure on the British government had resulted in a curtail-
ment of Canadian pelagic sealing, but that only served to encourage Japanese sealers to exploit the resource. The results were devastating—not only
to the numbers of seals, but to the natural balance of the herd. Adult bull
fur seals, many weighing five or six hundred pounds, had enough body fat
to weather the harsh winters on the Bering Sea, but the much smaller cows,
many under a hundred pounds, and adolescent males could not survive
the winters and had to migrate—some as far as three thousand miles—to
warmer waters along the western coast of Canada and the United States.
As a result, during the spring return migration, sealers killed a dispropor-
tionate number of females and young males. When females did reach the
rookeries, the older males became unusually violent as they competed for
mates—occasionally even killing the already dangerously scarce females in
the fray.[48]

Even with extensive research evidence to the contrary, the government
had been advised by Henry W. Elliott, a scientific illustrator–explorer who
did occasional work for the U.S. National Museum, that the North American Commercial Company's land hunting was to blame for the continuing decline of the population, not pelagic sealing. Since the 1896 Fur Seal

Commission, scientific investigations had determined otherwise, and now the FSAB, too, disagreed with Elliott's assessment. The rate of decline was noticeably more rapid on St. Paul than on St. George, though the land hunting practices were the same on both islands. "The difference," the board concluded, "seems to be due to the effects of Japanese sealing."[49]

In November 1909, David Starr Jordan submitted the FSAB's report to Secretary of Commerce and Labor Charles Nagel. It advised the United States to convene a conference of biologists and diplomats from each of the four major sealing nations with the goal of outlawing pelagic sealing internationally. The board also recommended allowing the continuance of the North American Commercial Company's lease and its permit to hunt on land; however, to ensure proper management, it proposed that the herd be regulated by the federal government, not the lessee. It wanted a government agent, assisted by the chief naturalist, administering the lease and establishing wise management policies and quotas for the annual killing of surplus males. Old bull seals were superfluous, it argued, because they were past breeding age but still competed with younger virile males. Furthermore, because fur seals were polygamous and breeding males mated with large harems of up to forty females, the reduction of male numbers was necessary to reestablish the balance of the herd.[50]

But Hornaday, now director of the Bronx Zoo and the country's leading wildlife advocate, denounced the recommendation. Despite his utter lack of familiarity with the Pribilof Islands and his long friendship with both Lucas and Townsend, Hornaday believed that the board was acceding to political pressure from the North American Commercial Company; he interpreted the language of herd management as an excuse to continue land sealing. To combat the recommendation, Hornaday enlisted the help of the Camp Fire Club of America, an organization of naturalists and game hunters, in mounting a public campaign. As chairman of the club's Committee on Wildlife Protection and a member of its board of governors, Hornaday lobbied Congress "to enact legislation calculated to save the fur seal species, and the fur seal industry."[51] In February 1910, Hornaday convinced Senator Joseph M. Dixon of Montana, who greatly respected Hornaday for his work in saving the state's bison herd from extinction, to hold hearings on the fur seals. Hornaday appeared before the committee to argue against renewal of the Pribilof lease. To his mind, such a renewal "would amount to the practical extermination of the herd."[52]

Hornaday based his conclusions on the field observations of Elliott, who worked on and off for various government agencies in Washington. Elliott moved freely across government institutional boundaries, which were easily permeated in nineteenth-century Washington. While his institutional affiliation was fluid, his focus had remained singular for more than three decades: since 1872, he had labored tirelessly, but without any scientific training or expertise, to save the fur seal. He was greatly disliked by the members of the FSAB and had in fact been excluded from participation on the board.[53] On the basis of his own incomplete observations, and a personal bias against the North American Commercial Company, Elliott had argued since 1890 that land hunting, not pelagic sealing, was the primary cause of the depletion of the herd. On Elliott's advice, Hornaday recommended that the Dixon Committee impose a ten-year moratorium on land sealing.[54]

The members of the FSAB were incensed. The 1896 Fur Seal Commission had already addressed Elliott's claim about land hunting and had concluded that there was no scientific data to support it. In the findings of that commission, Frederic Lucas had explained that land hunting "under ordinary circumstances would be wholly unjustifiable," but the "justification for this close killing is found in the existence of pelagic sealing, which spares nothing, and renders it proper and desirable."[55] For more than twenty years, the scientists of the FSAB had been collecting data to inform the best way to reach their stated goal of preserving and rehabilitating the fur seal herds. For Hornaday to question the FSAB's recommendations was to question the validity of its science.

David Starr Jordan believed that Hornaday's preservationist stance was unreasonable and was based on emotion rather than science; after all, Hornaday had never so much as seen a fur seal in the wild. Jordan and the other members of the FSAB concluded that the natural balance of the herd had been so disrupted by human interference that it could not easily right itself and would have to be managed. In May 1910, Secretary Nagel sided with Jordan and the FSAB and authorized the advisory board's plan, giving the lessee the go-ahead to harvest some twelve thousand young male seals.[56]

When Hornaday learned that Nagel had renewed the lease, he charged the secretary with violating the public trust. Nagel responded with a clumsy, anthropomorphic statement in the *Washington Post*: "If all the males are allowed to live, there wouldn't be enough wives in the Pribilofs to go round. . . . The male seals fight over the female seals. As a result, the casualties are enormous, and would, in the course of time, work to the extermination of all the fur-bearing animals on the islands."[57] Nagel was a corporate attorney, not a scientist, and his explanation was an unfortunate exaggeration of the facts that did little to persuade the American public that he was working in the best interest of the fur seals.

Hornaday, by contrast, was the public's trusted expert on the protection of wild game—a twenty-year veteran spokesman for conservation and a practiced popularizer of scientific data. He responded publicly to Nagel's awkward statement with loaded protectionist rhetoric that was calculated to escalate public outrage. Privately, Hornaday sent a damning letter to Nagel:

> Why did President Taft send a special message to Congress to provide against the making of a new killing lease? To stop the killing of the fur seals on the Pribilof Islands. Did the President, or did Senator Dixon's committee, or the United States Senate, intend for one moment that you should go right on in the bloody killing business without a halt? No! A thousand times no, and you know it![58]

For Hornaday and Elliott, the dispute was no longer a simple matter of conflicting viewpoints over wildlife management policy; they were prepared to raise the stakes.

Once the sanctioned culling of 12,920 fur seals in the summer of 1910 was over, Hornaday declared "open war . . . on all persons responsible for that treacherous slaughter." He and Elliott quickly embarked on a furious public campaign "to fight to the absolute defeat of one side or the other."[59] Together, they made the unsubstantiated claim that not only had superfluous males been slaughtered during that year's harvest, but that Nagel and the FSAB alike had permitted the hunters to "stab" and "club" pups, females, and "illegal yearlings," leaving a breeding population of "weaklings."[60]

In the months that followed, the *New York Times* was Hornaday's veritable mouthpiece, publishing almost daily accounts of the controversy along with anonymous editorials clearly written by Hornaday and Elliott. Lucas responded to Hornaday's attacks in signed editorials addressed to the *Times*. He also joined Townsend and other members of the FSAB in writing longer pieces intended for academics and an educated public in the semipopular *Forest and Stream*, which was edited by the conservationist George Bird Grinnell, and in *Science*, the journal of the American Association for the Advancement of Science. They argued that Hornaday had no knowledge of the Pribilofs and, moreover, did not understand the science that justified and even required management of the fur seal herd.

Apart from the rancor it inspired, the fur seal controversy revealed real ideological differences between Hornaday and his old friends Lucas and Townsend. The triumvirate had split along the ever-widening chasm in the American conservation movement between preservationists and conservationists.

Hornaday embodied the protectionist point of view, believing that species should be saved for posterity, while Lucas, Townsend, and others on the Fur Seal Commission believed that the herd could be managed as a natural resource. In the end, it can be said that they were all fighting for the same result—to save another species from extinction—but their mutual distrust grew corrosive over time.

RETURN OF THE ELEPHANT SEAL

In the midst of Hornaday's war, Charles Townsend was asked to serve as acting director of the American Museum of Natural History, after Hermon C. Bumpus resigned his position in 1910. Townsend accepted and was temporarily released from his duties at the New York Aquarium. As virtually his first order of business, Townsend assigned the preparator Frederick Blaschke to mount a group of fur seals from skins collected two years earlier on the Pribilof Islands.

The mounted group had only recently been installed when Townsend received incredible news about another seal species thought to be extinct. Northern elephant seals had been spotted on Guadalupe Island, off the west coast of Mexico's Baja California peninsula. In 1892, Townsend was thought to have killed the last remaining individuals of the species at Guadalupe Island, while on a collecting trip for the U.S. Fish Commission.[61] Thrilled by the possibility that nature had persevered, and eager to see for himself, Townsend quickly organized an expedition, jointly sponsored by the AMNH, the New York Zoological Society, and the Bureau of Fisheries (formerly the U.S. Fish Commission), which allowed it to use the steamer *Albatross*. Townsend hoped to collect four northern elephant seals for a new museum exhibit and several live specimens, if possible, for the aquarium.[62]

The *Albatross* arrived at Guadalupe Island on the morning of March 2, 1911. Townsend sent the scientific staff ashore to begin the day's collecting at a deserted camp east of the volcanic island's northern point, while he remained on board to search for the site of the old elephant seal rookery on its northwestern side. Once in position, the *Albatross* anchored just off the point, and Townsend and the ship's crew, keeping just outside the breakers, rowed a small boat along Guadalupe's northwestern coast toward Steamer Point, where he had last collected northern elephant seals nearly two decades earlier. There again, Townsend found a group of elephant seals, about 125 in all, lying on the sand beach below the high cliffs, protected on either side by large rockslides. He ordered the crew closer and killed two large seals, one bull and one cow, and returned to the *Albatross*.[63]

Fig. 5.8. Charles H. Townsend's photograph and field study of the bull elephant seal collected and mounted for the U.S. National Museum. (© Wildlife Conservation Society. Reproduced by permission of the WCS Archives.)

After offloading the skins, Townsend ordered the ship back to the east side to pick up the scientific staff, while he returned to the beach with the remaining crew, armed with larger boats and nets. He spent the afternoon photographing the seals in a variety of natural attitudes, although he admitted that some of the more aggressive expressions were induced. "I focused my camera on an elephant seal at a distance of eight to ten feet," he wrote, "and then had a sailor kick the animal violently in the ribs."[64] Another was so docile that a sailor had to climb on its back before it would assume a fighting attitude. After several hours, he obtained "about fifty good negatives."[65]

Townsend then set about collecting six live yearlings. He reported that the young seals were wound up in nets "so tightly that we could handle them like bales."[66] They were kept this way "to prevent them from biting, or escaping from the boats."[67] At dusk, the *Albatross* returned and anchored for the night about half a mile offshore.[68]

The next morning, the sea was too choppy to go ashore in open boats, so the crew spent the day preparing the two large skins, scraping them clean and salting them, so that they would be well preserved for transport to the AMNH.[69] The old male was sixteen feet long, with a proboscis as long as its head, and its skin, Townsend remembered, was extremely heavy, even fully flensed. They packed it "in a full sized barrel which it completely filled and that without the skull."[70]

On the morning of March 4, Townsend took a crew ashore and shot two more bulls, equal in size to the one collected two days earlier. It took several experienced men all morning and half the afternoon to skin and skeletonize the unwieldy specimens in the lee of a high, concave cliff with a sheer face more than two thousand feet tall.[71] Townsend recalled:

> Our knives dulled so rapidly in skinning them that it was found neces-
> sary to have a grindstone sent ashore and to keep two men busy at the
> task of sharpening. The carcasses were so heavy that it required all the
> strength of half a dozen men to turn them over, with the aid of a rope and
> hand-holes cut in the skin.[72]

One member of the scientific crew, the ornithologist Pingree L. Osburn, later remembered that the hard work was further slowed "by loose flying boulders from the top of the cliff."[73] As darkness fell, the skins were taken aboard, and the *Albatross* turned toward San Diego, "in order that the young elephant seals and the large skins might be shipped eastward without delay."[74]

The entire next day was occupied by cleaning the skins and skeletons of the four specimens and preparing them for shipment. When they arrived in San Diego on the morning of March 6, the six yearlings were crated individually in enormous iron water tanks, specially constructed to transport them, and because they were refusing to eat, shipped by express train to the New York Aquarium. For six days, the seals were without food, but when uncrated at the aquarium, they appeared to be in good condition.[75] Days after their arrival, they were swimming around the aquarium pool and eating live fish tossed in to them. Townsend told *Forest and Stream*, "Our success at Guadalupe Island was quite beyond expectation."[76]

For close to six months, the elephant seals thrived—and Townsend even sent two, nicknamed Jim and Bob, to the National Zoo in Washington. Shortly thereafter, however, the four yearlings at the New York Aquarium died. Jim and Bob survived to their second birthday—marked by zoo director Frank Baker on February 4, 1912, with an extra bucket of fish—but neither would live out the year.

Hornaday could not resist commenting on the perceived irony of Townsend collecting specimens of a species that for nearly three decades was thought to be extinct, even though he had done the same when, in 1886, he had collected a live bison for the U.S. National Museum. Hornaday cautioned

Fig. 5.9. Detail of the group of elephant seals collected by Charles H. Townsend. The bull pictured here is still on display at the AMNH. (Image #38717, American Museum of Natural History Library)

that Townsend's expedition may have unwittingly revealed the location of the northern elephant seals to unscrupulous hunters who would soon "go to those islands and 'clean up' all the remainder of those wonderful seals."[77] One step ahead of Hornaday, Townsend had a plan to discourage such hunters. Always working from within the government system, he announced that a joint action undertaken by U.S. and Mexican authorities would be carried out to protect the elephant seals.[78] But Hornaday was unconvinced that any amount of legal protection could prevent the extermination of the herd. "One hunting party could land on Guadalupe and in one week totally destroy the last remnant of this almost extinct species," he wrote. "To-day the only question is, Who will be mean enough to do it?"[79]

Townsend took some consolation from the fact that, even though all six of his live animals had died and the species itself teetered on the brink of extinction, the skins and skeletons he had collected on Guadalupe Island would live on in American natural history museums. He declared:

The completion of a group of elephant seals in the American Museum of Natural History in New York, mounted according to photographs and actual measurements, will soon give us a graphic view of this large and remarkable North American animal that came so near to being lost to science.[80]

The carcasses of Jim and Bob were donated to the U.S. National Museum. The four yearlings that died at the aquarium were divided between the AMNH and the Brooklyn Museum.[81] It would take until 1920 for the American Museum to complete its group of six elephant seals, at which time they were temporarily exhibited in a case in the central section of the museum.[82] In 1922, when the AMNH opened its new Hall of Ocean Life, the elephant seal group found a permanent home. Aptly, it was the same year that the Mexican government, at Townsend's urging, finally passed protective legislation that banned hunting of the northern elephant seal and made Guadalupe Island a biological reserve. A few years later, when the population expanded into Southern California, the U.S. government passed similar legislation, which was strengthened in 1972 when the northern elephant seal was included in the Marine Mammal Protection Act.

THE INTERNATIONAL BAN ON PELAGIC SEALING

While Townsend was en route to Guadalupe Island, Secretary of State Philander C. Knox, aided by Secretary Nagel, successfully negotiated a treaty between Great Britain and the United States to end pelagic sealing in the North Pacific—to go into effect if Japan and Russia joined and ratified it.[83] The treaty provided that if Canada and Japan would agree to cease pelagic sealing, in return, the United States and Russia, for the next fifteen years, would pay to each country a percentage of their receipts from land hunting. Under mounting public pressure to save the fur seals, the United States, after thirty years of failed negotiations, was now forced to concede remunerations to Canada and Japan.[84]

As the international negotiations to end pelagic sealing rapidly progressed, both sides of the seal controversy found enough common ground to support Nagel's work. Even Townsend and Hornaday found reason to work together on behalf of their longtime friend and colleague Frederic Lucas. They successfully lobbied AMNH president Henry Fairfield Osborn to hire Lucas as the museum's next director, and on May 9, 1911, the *New York Times* announced that Lucas would be the new permanent director of the AMNH. Osborn explained that he had hired Lucas because "his long experience in Ward's

Natural Science Establishment, in the United States National Museum and as Curator-in Chief of the Brooklyn Institute Museum eminently qualified him for the office."[85] The *Times* reported that Lucas "comes to the Museum with strong indorsements" from both Hornaday and Townsend, noting that the three were "classmates thirty-one years ago in Ward's Natural Science Establishment at Rochester," and that the museum's trustees believed "the close co-operation which will result among the three great institutions dealing with natural history will benefit all of them materially."[86]

As Hornaday, Townsend, and Lucas, for the moment, came together, so did the principal nations that had so long contended over the right to hunt fur seals in the Bering Sea. In the same month that Lucas assumed the directorship of the American Museum, delegates representing Great Britain, Russia, Japan, and the United States convened in Washington to attend an international conference "to frame a treaty for the protection of fur seals, plumage birds, sea otter, and other sea animals."[87] On July 7, 1911, delegates from all four nations signed a treaty prohibiting, for a period of fifteen years, pelagic hunting of fur seals and sea otters in the North Pacific. Before the treaty could go into effect, however, it had to be ratified by the governments of each nation. In the United States, in addition to ratification by the Senate, Congress would have to enact legislation to ensure that the terms of the treaty became law.[88]

While Congress considered the treaty and worked to carry out its provisions, Hornaday seized the opportunity to attach a clause to prohibit the killing of fur seals on land, renewing his argument with Townsend and Lucas. Before the close of the congressional session, a resolution was introduced in the House of Representatives to suspend for fifteen years the killing of fur seals at sea and on land. Townsend spoke out against the resolution in a paper he read at the forty-first annual meeting of the American Fisheries Society, which was later published in the October 27, 1911, issue of *Science*:

> While a cessation of land killing for a season or two might cause no serious trouble, the fifteen-year period specified is not only too long, but positively dangerous, as the Bureau of Fisheries would be powerless to apply the necessary remedy for the evil of overcrowding by males when it becomes serious.

Townsend disparaged the resolution as the work of "men who have not been to the islands for twenty years" and "men who have not been there at all, and whose opinions upon the subject are of little value."[89] Although he did not name names, the former rebuke clearly referred to Elliott, and the latter to

Hornaday and the Camp Fire Club of America. Despite the research amassed by scientists against the advisability of the closed season, on August 24, 1912, Congress passed a law enacting the requirements of the fur seal treaty of 1911 with a clause prohibiting all killing on land for a period of five years.[90]

By February 1912, the defeated scientists had renewed the debate with great fervor. Lucas published an editorial in the *New York Times*—again struggling to explain the need to reduce the number of male fur seals. By now his frustration was beginning to show, as he wondered why the public could not understand the problems that a small ratio of males to females would present, when "the [management] plan is practiced with so much success with all of our domesticated animals." Like Townsend, he questioned how "those who have never visited the Pribilof Islands are so much better informed as to the proper method to be pursued than are the English, Canadian, and American naturalists who have been on the islands and have actually studied the fur seal question."[91] In an unsigned response in a neighboring column, probably penned by Hornaday, the *Times* dismissed Lucas's comparison to domestic animals. "The killers select the best of the 'bachelors,' they leave the worst," the writer argued. "In the case of domestic animals the worst and weakest are killed, the best are left to perpetuate the species."[92]

That spring, Townsend and Marshall McLean, a general practice lawyer and member of Hornaday's Committee on Game Protection of the Campfire Club of America who had also participated in the 1910 Senate hearings, aired their differences in *Science*. McLean entered the fray in defense of Hornaday and the Camp Fire Club as a result of the accusations that Townsend made in the paper he read at the annual meeting of the American Fisheries Society.[93] Townsend had hailed the end of pelagic sealing and the international convention that made it possible. He had summarized the long controversy and the work of the scientists appointed to the international fur seal commissions, whose research had resulted in the fur seal being the most studied and best understood mammalian species. With the end of pelagic sealing, Townsend believed that these scientists now deserved the opportunity to "apply scientific methods to the rehabilitation of the small herd remaining on the Pribilofs." He had detailed the possible management strategies and the supporting science, including the present responsibility to manage the herd's superfluous males.

In the February 2, 1912, issue of *Science*, McLean argued against Townsend's suggested methods for herd management, especially the annual killing of superfluous males, by repeating Hornaday's testimony from the Senate hearings of 1910. McLean also maintained that his reasoning was not based "on the conflicting reports of scientists, but on the broad principle that when a species of wild life has become so depleted as to be in danger of extinction, the best remedy is to let it absolutely alone." McLean must have realized that, given the scientific community's apparent consensus and the fact that the preservationist stance unfortunately hinged on Elliott's long-disproved "science," he would have to shift the debate.[94] Doing so would be challenging, given that Hornaday and Elliott had engaged in battle with the most respected scientists of the time, who were intent on defending their research.

In the March 1, 1912, issue of *Science*, responses came both from Townsend and from George Archibald Clark of Stanford University, who had also served on the fur seal commissions with David Starr Jordan. In separate editorials, Townsend and Clark pointed to the fact that fur seals were not like other wild animals, but were instead a valuable economic resource that should be managed like the fisheries of the United States.[95] Clark further noted that seals were polygamous and therefore should not be compared with "pairing animals like the deer, bear, duck, or quail. . . . Its true analogies are with the domestic animals—cattle, horses, sheep, poultry."[96] Lucas reinforced this point in a brief article in *American Museum Journal*, the popular publication of the AMNH. "The regulated killing of young males on land has caused no decrease in the fur seal herd any more than the systematic killing of cattle and sheep depletes the stock-raiser's herds and flocks."[97]

Ironically, Lucas worried that the museum's group of fur seals, mounted at Townsend's direction in 1910–1911, might undermine the public's understanding of the violent behavior between adult bulls and bachelor seals by depicting two specimens close together and in apparent harmony. "Our fur seal group in the Museum is necessarily untrue to nature," Lucas lamented, because the bachelors' natural behavior dictates that they "keep by themselves, approaching the breeding grounds, or rookeries, literally at the peril of their lives." However, he defended the yearling placed at the front right of the group, explaining that it "belongs there, because yearlings are permitted to come to the edge of the rookeries where they play with the baby seals or pups." He indicated that the unfortunate error was a result of "endeavoring to give a comprehensive idea of the fur seals"—that is, of both sexes, and at various ages.[98]

In the coming years, Lucas would insist on higher and higher levels of scientific accuracy for new groups mounted for the American Museum of Natural History—and he would find unmatched support in this enterprise from Henry Fairfield Osborn. As president of the board of trustees, Osborn found it "gratifying to realize that the American Museum of Natural History has held from the first a position as one of the centers of the conservation movement." This position was achieved not only by advancing science's understanding of species threatened with extinction, but more importantly, and perhaps more effectively, by promoting species conservation through popular instruction:

> Among the hundreds of thousands who annually pass through the institution's halls are many who gain knowledge and an abiding interest in nature, the very mainsprings of the conservation idea. It is cause for congratulation also that the Museum's influence for the preservation of animal life is continually increasing as advances are made in methods of exhibition and public education.[99]

Fig. 5.10. Charles H. Townsend (*third from right*) aboard the *Nourmahal* on the William Vincent Astor Expedition to the Galápagos Islands in 1930. Townsend continued collecting specimens for conservation research under the auspices of the New York Zoological Society (now Wildlife Conservation Society) throughout the 1930s.
(© Wildlife Conservation Society. Reproduced by permission of the WCS Archives.)

With Lucas as its director, the AMNH would soon begin the most important period of its history—as an innovator of museum display and popular
instruction, and as a global conservation organization.

AT LAST VINDICATED

As a result of the ongoing and "sharp controversy" surrounding the fur seals,
the secretary of commerce in 1914 appointed scientific administrators "dispassionate," "unprejudiced," and "free from all previous connection with
the subject" to carry out the law passed in August 1912, calling for a closed
season on the Pribilof herd. After receiving nominations, the secretary selected Wilfred H. Osgood, assistant curator of mammalogy and ornithology
at the Field Museum of Natural History; naturalist Edward A. Preble of the
Bureau of Biological Survey; and George H. Parker, professor of zoology at
Harvard University. That summer, the three scientists traveled north to the
Bering Sea and the Pribilof Islands. Their two months of observing the herds
culminated in a lengthy report, in which the three biologists made recommendations "previously urged, some of them repeatedly"—including, most
significantly, the recommendation to amend the law of 1912 to allow the
killing of bachelor males on land. On the basis of the population studies
it had conducted, this new scientific committee determined that the high
numbers of bachelor seals observed had resulted from the cessation of
land hunting, and that the dangerously low number of females could have
resulted only from decades of pelagic sealing. Townsend, Lucas, and the
scientific community at last were vindicated.[100]

However, by the time the report was released to the public in 1916, the
New York Zoological Society and Charles Scribner's Sons had published
William T. Hornaday's preservationist tract, *Our Vanishing Wild Life*, in
which Hornaday claimed credit for himself and Henry Elliott for saving the
fur seals. Osborn wrote in the book's foreword that the NYZS, in cooperation with "many other organizations in this great movement, sends forth
this work in the belief that there is no one who is more ardently devoted
to the great cause or rendering more effective service in it than William T.
Hornaday."[101] Osborn believed, too, that the book was "destined" to "exert
a world-wide influence" and would "arouse the defenders and lovers of our
vanishing animal life before it is too late." But once again, Hornaday seized
the opportunity to condemn academics and museum administrators for doing too little in service of conservation, arguing that when these institutions and their scientists "talk about living things, the public listens with
respectful attention."[102]

In dividing institutions according to their contributions to conservation, Hornaday claimed that the American Museum's administrators had acted responsibly but had "room for improvement." Among those he believed could do considerably better were his former bosses at the U.S. National Museum and several institutions that now employed the next generation of taxidermists from Ward's, most notably the Field Museum and the Carnegie Museum. Finally, he singled out the National Zoo, the institution he had angrily abandoned in its infancy, as having "done nothing noteworthy in promoting the preservation and increase of the Wild life of America."[103] The only individuals who earned Hornaday's praise were Henry L. Ward at the Milwaukee Public Museum, the son of his mentor; Lewis L. Dyche at the University of Kansas, his former apprentice; and Joseph Grinnell, the first director of the Museum of Vertebrate Zoology at the University of California.[104]

Not surprisingly, the museum administrators did not share Hornaday's dim view of their contributions. At the annual meeting of the American Association of Museums in 1915, they took the opportunity to respond publicly. Alja Robinson Crook, curator of the Illinois State Museum, presented a paper titled "The Museum and the Conservation Movement," in which he directly countered Hornaday's attack: "Museums from the beginning have preached conservation of natural resources. . . . It may be said that they were the earliest and are now the most consistent representatives of the idea though the fact may not be generally appreciated." The prominence conservation had attained in recent years, he asserted, would not have been gained as swiftly were it not for "the long quiet work of museums in this line." Much of that work had been carried out, he argued, by men like Hornaday, who had created "special exhibits" to represent "the conservation idea." The new conservation movement, he further contended, "was born in the minds of men who were influenced by the facts so well shown in museums."[105]

Crook encouraged museums to continue this work by creating new exhibits that would more explicitly advocate conservation ideals. As an example, he suggested the "passing procession of animals marching into oblivion":

Here may be grouped the larger mammals, the birds, and fishes—those which are dwindling, those that have recently disappeared (within the last fifty or one hundred years), and those that became extinct in quaternary, tertiary, cretaceous, or preceding geological periods. . . . No portion of the conservation exhibit is more attractive than this since probably a larger number of visitors in the ordinary museums are interested in animals and in the vistas of the geological record than in any other department of natural history. I am glad to note that a number of museums have been devoting considerable attention to this work.[106]

He offered other possible conservation topics, including forest management, mining, soil conservation, and "waste in gas and oil production." He concluded by challenging Hornaday's notion that an epic battle was the only way to save America's endangered wildlife, suggesting instead that over time, museums could be "the most effective of all conservation agents," not by persuading lawmakers directly, but by educating the general public about "the wrong attitude of man towards nature, and by pointing the way to good practices."[107]

Little did anyone suspect that Carl Akeley, then known only for his accomplishments as a taxidermist, was about to bring together the best of both movements—simultaneously working to educate museum visitors about wildlife and their habitats through exhibits grander than ever before conceived, and effectively convincing heads of state that conservation was in their economic and political interests. Hornaday, Lucas, and Townsend, old friends and adversaries, had laid the groundwork for a new environmental ethic that stressed the fragile balance between humans and nature—but Akeley, standing on the shoulders of his mentors and with a singular vision for the future, was about to embark on the most ambitious and significant project in the history of American natural history museums.

CHAPTER SIX

"Brightest Africa":
Carl Akeley and the American Museum's Race to Bring Africa to America

I have always been convinced that the new methods of taxidermy are not being used to the full; that, although the taxidermic process has been raised to an artistic plane, a great opportunity still remains for its more comprehensive use in the creation of a great masterpiece of museum exhibition.
—Carl E. Akeley[1]

Upon completing his third African expedition, Carl Akeley returned to Chicago in November 1911 long enough to make the necessary arrangements to move to New York and assume his new position as chief taxidermist of the American Museum of Natural History. Akeley was excited for the future and his mind was filled with dreams. He hoped to bring together his improved method of taxidermy and his idea of curved-back dioramas to put in "permanent and artistic form a satisfying record of fast-disappearing fauna and give a comprehensive view of the topography" of the African continent.[2] He admitted to the *Chicago Daily Tribune* that, though he had several designs in mind, the final design for such an African hall was still taking shape. "Perhaps I will have the elephants at feed in a native shamba or garden," Akeley told the *Tribune* reporter, though he rued that this "would require an acre of space."[3] When he arrived in New York in January to present the report of his trip to the New York Zoological Society in the ballroom of the Waldorf-Astoria, he found that he had a group of willing supporters. Akeley was preceded on the program by reports from William T. Hornaday, director of the Bronx Zoo, and Charles H. Townsend, director of the New York Aquarium.

Hornaday chronicled his continued work in the protection of wildlife, including his recent failed attempt to set reasonable bag limits for wild fowl in New York State. A bill he had supported before the state legislature had been successfully lobbied against by the newly founded American Game Protective and Propagation Association. Hornaday pointed out that the group, which supposedly advocated harmony between hunters and conservationists, was actually funded by "the heavily capitalized makers of repeating guns and ammunition" and that the founders included the manufacturers of "two automatic and three 'pump'-gun slaughtering machines" used by commercial hunters.[4] The defeat of this legislation was sadly predictable.

Hornaday had just published *Our Vanishing Wild Life*. Opposite the book's title page was a photograph of the last living passenger pigeon, housed at the Cincinnati Zoo, and a quoted passage from the 1857 report of a select committee of the Ohio State Senate recommending against protection of the bird, which stated that "no ordinary destruction can lessen them."[5] Hornaday argued that "the folly of 1857" was "the lesson of 1912," demonstrating that "any wild bird or mammal species can be exterminated by commercial interests in twenty years time, or less."[6] He praised what he saw as the more enlightened state of Pennsylvania, whose governor had declared a statewide "Bird Day" for the coming spring, and Pittsburgh, which had named Frederic S. Webster as its first city ornithologist: "The duty of the new officer is to protect all birds in the city from all kinds of molestation, especially when nesting; to erect bird-houses, provide food for wild birds, on a large scale, and report annually upon the increase or decrease of feathered residents and visitors."[7]

Following Hornaday, Townsend described his recent expedition aboard the *Albatross*, under the joint auspices of the New York Aquarium and the American Museum. With a series of lantern slides, he showed the audience the herd of northern elephant seals—thought to have been extinct—that he had recently observed on Guadalupe Island, off Baja California.

Finally, Akeley rose to the podium to describe his travels to Africa and his plans for an unprecedented hall of African wildlife, with a group of elephants as its centerpiece.[8] Akeley believed that the moment was right:

> Twenty-five years ago, with innumerable specimens at hand, its development would have been an impossibility. Even if a man had had all the animals he wanted from Africa, he could not have made an exhibit of them that would have been either scientific, natural, artistic, or satisfying, for twenty-five years ago the art of taxidermy and of museum exposition of animal life hardly existed. Likewise, in those days much of the information that we had about animals through the tales of explorers, collectors, and other would-be heroes was ninety-five per cent inaccurate.[9]

Now, Akeley had developed the necessary methods to display the specimens he had collected, and through a collaborative effort with the scientists who were gathering new data about the animals' habitats and behavior, the museum would have the most scientifically accurate exhibit of African wildlife anywhere in the world. But time was running short. He didn't believe that elephants were at risk of imminent extinction—though he reported that "the big bulls" were steadily being "cleaned out" by ivory hunters[10]—but other species were in a more precarious position. Just as any African hall undertaken twenty-five years earlier would have been unsatisfactory, Akeley feared that "twenty-five years hence the development of such a hall will be equally impossible for the African animals are so rapidly becoming extinct that the proper specimens will not then be available."[11] In fact, he predicted that "by the time the groups are in place in African Hall, some of the species will have disappeared. Naturalists and scientists two hundred years from now will find there the only existent record of some of the animals which to-day we are able to photograph and to study in the forest environment."[12]

After his lecture, Akeley met yet another ally whose support would prove especially important in the pursuit of his vision: Frederic Lucas, the new director of the American Museum. Hermon C. Bumpus had left the post six months earlier, while Akeley was in Africa, to assume the presidency of the University of Wisconsin. Lucas had vocally championed Akeley's work at the Field Museum, but later Akeley fondly remembered that this first meeting was the beginning of their "delightful association."[13]

Lucas understood—and was able to convince Osborn of—Akeley's larger vision for a hall of African wildlife. The group of elephants Osborn commissioned could serve as the first phase and centerpiece of the new exhibit, but the elephants were not unique in their status as a rare and possibly endangered species. Another AMNH expedition to the Belgian Congo, led by Herbert Lang and James Chapin in 1911, had collected the first known specimens of the okapi,[14] and Frederick Blaschke, who had studied sculpting under Rodin in Paris and trained in Akeley's method at the museum, was already at work on a pair of zebras and a hippopotamus acquired from the Central Park Zoo. Akeley's notion of an entirely new hall would allow these new groups to be mounted by a small group of preparators together and presented in a single, harmonious design.

As the new director, Lucas also must have felt the pressure of competition, as Childs Frick, on behalf of the Carnegie Museum, had been in Africa collecting at the same time that Akeley and Theodore Roosevelt were there—and groups from that expedition were already beginning to appear in the Carnegie's new building. In April 1911, the CMNH unveiled a group of oryx and a pair of giraffes, mounted by Remi and Joseph Santens. In June of the previous year, William Hornaday had invited the Santens brothers to come to the Bronx Zoo, where they could observe live animals and make studies in clay of the giraffes and oryx they intended to mount for the Carnegie Museum.[15] This pioneering practice would soon become a typical method of study for museum taxidermists, who did not do their own field work and thus did not have the opportunity to view live animals in the wild, but who were expected to produce scientifically accurate mounts. Once those groups were installed, the Santens brothers turned immediately to mounting groups of white-bearded wildebeest and zebras. By the time Akeley arrived in New York in early 1912, the Santens brothers were just weeks from completing their wildebeest group, and their zebra group—praised by Hornaday as "a beautiful and spirited achievement"—would be installed before the end of the year.[16]

To outdo the Carnegie Museum of Natural History, however, would require considerable funding. With the support of Lucas and Osborn, the AMNH board approved Akeley's plan for a hall of African wildlife, but they stipulated that he would first have to complete a portion of the exhibits, and they would then seek further assistance from the city. The work would have to be swift, efficient, and attractive to potential donors. There was no time to waste.

DESIGNING THE HALL OF AFRICAN MAMMALS

Akeley's plan for a hall of African wildlife was ambitious and unlike anything thus far attempted in a natural history museum. The overall design would represent the range of ecosystems in Africa, "from the Mediterranean on the north to the Tableland Mountain at Cape Town and from the east coast to the west coast," complete with flora and fauna.[17] Akeley, a skillful and prolific sculptor, also hoped to incorporate his life-sized bronze sculptures of Nandi lion hunters. The hall would measure 60 by 152 feet. The main floor would rise 17 feet to a gallery level, and 30 feet from its center to the ceiling. In the center of the hall, on the main floor, would be an unenclosed exhibit of a family group of four elephants flanked on either end by two groups of rhinoceroses, one of the black species and another of the white. Akeley intended to present this central exhibit "in statuesque fashion."

For Akeley, the choice of the elephant as the main feature of the hall was justified, as it was "typical of Africa" and was the world's largest land mammal. Akeley titled the elephant group "The Alarm" and described its intended narrative:

> The composition . . . shows the bull scenting danger, silently feeling for scent with his trunk, ears fully extended to catch the least sound, for he does not see the source of disturbance. The attitude of the cow indicates that she has seen the intruder and has "frozen," ears back, trunk pendant, prepared for any move she may decide on, whether attack or retreat. The calf, conscious of the alarm, is snuggling up to its mother for protection. The young bull to the right, startled, has started forward to swing around and face the danger, his trunk thrown back to catch the scent, and his ears forward to catch the sounds.[18]

The central exhibit would be surrounded on all sides by groups making up "typical" African scenes. The cases would recede into the wall, creating "a sort of annex which will not encroach upon the measurements of the hall proper." There would be forty of these cases: twenty on the first floor and twenty smaller cases on the gallery level. Akeley described the intended effect as "looking out through open windows into an African out of doors." To enhance the effect, Akeley designed the groups with panoramic backgrounds, which he intended to be painted "by the best artists available and from studies made [by the artists] in Africa." Each of the forty groups, in representing a specific habitat, would be a "composite—that is, as many species will be associated in each of the groups as is legitimate with scientific fact." He gave as an example one of the large main floor corner groups, which would represent

> a scene on the equatorial river Tana, showing perhaps all told twelve species in their natural surroundings with stories of the animals and a correct representation of the flora. In the foreground on a sandbar in the river will be a group of hippos; across the stream and merging into the painted background, a group of impala come down to water; in the trees and on the sandbars of the farther bank two species of monkeys common to the region; a crocodile and turtles basking in the sun near the hippos and a few characteristic birds in the trees.[19]

To protect the specimens and accessories on exhibit from exposure to sunlight, which did the greatest damage by fading natural fibers—especially hair and feathers—as well as wax foliage, Akeley planned new systems for limiting their exposure to light in the exhibit environment. The protection of specimens was a matter that concerned even the museum's administration. After noting that as a result of exposure to sunlight, "many of the mammals now on exhibition are worthless and some have been ruined in the short space of five years,"[20] Lucas instituted a program to replace the museum's clear glass windowpanes with ground glass that diffused light. He felt strongly that because museums were "leaders in the conservation movement," it was just as important to preserve specimens as it was to collect them.[21]

Because the museum's southeast wing, which would eventually house the African hall, had not yet been constructed, Akeley's overall design included his own ideas for mitigating the challenges of the exhibit environment. If the new building design was to include natural lighting, Akeley wanted to install automatic shutters on the hall's skylights to "maintain a uniform light." With the central hall dimly lighted, each of the "annex" cases would have a small amount of electric lighting from within and above. Lighting cases from within to lessen the problems of reflection, especially typical in four-sided glass cases, was an entirely new concept. To further reduce annoying glare, Akeley designed each case with its glass front slanted toward the floor to reflect only "the dark floor of the hall." To control humidity and dust, Akeley would install a monitored air filtration system for all of the cases. Akeley's vision for the African hall—with its receding cases, panoramic paintings, composite groups, and inventive and preservation-minded case designs—was nothing short of revolutionary. Akeley's achievement would far surpass the goals of the new taxidermy movement begun more than three decades earlier with the founding of the Society of American Taxidermists.

COMPETITION WITH THE CARNEGIE MUSEUM

While Akeley awaited final approval for his African hall plan at the American Museum, he continued, on contract, to mount African exhibits for the Field Museum, including ambitious dioramas of Cape buffalo and eland. James L. Clark recalled that Akeley mounted these two exhibits as a way "to-keep-the-pot-boiling" at the American Museum.[22] The Field Museum's African exhibits set the standard to achieve or surpass. The higher he set the bar by adding to the exhibits in Chicago, the more likely it was that the board in New York would have to approve his expensive plans for the American Museum—and soon the pressure was not coming only from Akeley.

With its talented taxidermists, Remi and Joseph Santens, the Carnegie Museum entered into the friendly rivalry. When Remi Santens joined the

museum in 1906, there were already several African specimens in the mammal collection, donated by Childs Frick—a native of Pittsburgh and a 1905
graduate of Princeton who later became curator of paleontology and a trustee
for the AMNH. Santens reported to CMNH director W. J. Holland that a third
of the Frick specimens were "not turning out satisfactorily in the tanning
process," because the "skins were not properly handled in the field." The
poor condition of the skins was the result of their having been transported
in a salt and alum bath, which other institutions had "entirely discarded as
an improper method of preparation." Before sending another expedition to
Africa, Santens recommended that Holland consult "Messers James Clark
and Carl Akeley both gentlemen being experienced men in the Collecting
of skins for mounting purposes," and both of whom he was certain would
"vouch the correctness" of dispensing with an alum bath transport.

In 1911, Childs Frick led another African expedition and again donated
specimens to the Carnegie Museum. Frick must have heeded Santens's advice on field preparation, as these specimens were quickly mounted. The
Santens brothers worked feverishly, mounting and designing family groups
of African mammals for the museum's Gallery of Mammals, where mammalian species were shown without regard to geography.

In the spring of 1913, the Carnegie Museum responded to Akeley's Cape
buffalo group with its own group of African buffalo mounted by Remi Santens. The scene was "a typical African papyrus marsh. The water and water
plants are as nearly true to nature as it is possible to make them." This
group represented a trial for a new technique developed by Remi Santens to
enhance the realism and scientific accuracy of the mounted specimens: the
buffalo had "on their bodies the mud and moisture of their natural habitat."
Even Santens was aware of the importance of his new technique. He wrote
to Holland:

> This (to the best of my knowledge) is the first instance wherein the spec-
> imen was placed on exhibition showing the natural signs of contact with
> his environment. It is liable to cause comment and criticism, but I fully
> believe it is a step in the right direction and will be followed by many of
> the leading taxidermists of the future.[23]

Santens's new technique was eventually adopted, though not on a large
scale.

Remi Santens, like Akeley, was constantly working on developing new
methods in taxidermy, and in June 1915, at the AAM meeting in San Francisco, he presented a new adjustable-frame mounting technique. He also dis-

cussed the limits on taxidermists of his generation and how they might be overcome. In his view, the greatest challenge was that "as a rule a taxidermist cannot collect his own animals." Because his experience making clay studies of animals at the Bronx Zoo had proved a success, he suggested that "in this case the modern method is to visit some zoological park where living animals of the species to be mounted can be studied."[24] His lecture was published in the *Proceedings* along with a photograph of the African buffalo group.

After Remi Santens returned from the AAM meeting in San Francisco, visiting several scientific institutions and museums along the way, he began mounting a nyala group, which he completed in January 1916. However, just days before the group was to be installed, a fire of "unknown origin" started in the taxidermy laboratory, just after one o'clock in the morning. Fortunately the night watchman quickly discovered the flames. Nevertheless, Remi estimated that the fire had consumed "fully of a year" of lab assistant Anna M. Dierdorf's artificial foliage. The "painted scenery" for the nyala exhibit was not fully replaced until July; soon after, in September, the group was installed and unveiled. Joseph Santens completed two additional African exhibits in October, a family group of gerenuk and another of dikdiks, but one year later he accepted a position as chief taxidermist for the Buffalo Society of Natural Science and left the Carnegie Museum.

Remi, now chief taxidermist, continued mounting African specimens, and in 1919 completed a pair of black rhinos (one collected by Childs Frick and the other by Theodore Roosevelt). For two months, beginning in early February 1920, Mrs. M. Clayton worked on the foliage for the group. By April, the accessories were completed, and the exhibit was unveiled on April 27, 1920. William Hornaday declared, "The Santens black rhinoceros group is truly a *tour de force*."[25]

"THE ALARM"

Facing pressure to compete with the Carnegie Museum and the Field Museum, American Museum president Henry Fairfield Osborn announced in 1913 that as the Congo expedition was "drawing to a close," the museum could now give "special attention" to the exhibition of the African mammals collected.[26] "The Museum is fortunate," he wrote, "in having secured Carl E. Akeley, an eminent and skillful animal sculptor and preparator, to direct this work."[27] Lucas arranged to have the old hall of North American mammals—the second-floor Southeast Pavilion—cleared of exhibits so that Akeley could set up his taxidermy workshop there. Nicknamed the

"elephant studio," the room provided a large space to work on the impos-
ing group of pachyderms.[28] "Such work depends on just a few men who can
carry it out," Akeley said. "To find people who can do the work, men of fit
training and sense to carry it to the finish, that is the difficult matter."[29] He
committed the next five years not only to completing the work, but also
to founding a "taxidermy studio" that he hoped would "prove a training-
ground for young men of ability and marked aptitude for the work," just as
Ward's had been for him and many other museum taxidermists.[30]

To maintain continuity in exhibits and to make room for the existing
mounted specimens of African mammals "scattered through various halls,"
Lucas removed current exhibits from the large wall cases in the hall adjacent to the Southeast Pavilion, which at the time was Asiatic Hall. The
old African specimens filled one entire wall of cases at the south end, and
"a reproduction of Bushman rock paintings" was added to tie the exhibits
together. The hall was cramped and not designed for the large habitat groups
Akeley had in mind, but the space was only meant to house a temporary African hall while the administration raised funds to complete the Southeast
Wing of the museum building, intended for the proposed Theodore Roosevelt Memorial Hall and African hall. It was for this eventual permanent
space that Akeley designed the elephant group.

Little was known about modern African elephants when Akeley was
sent to collect them. It wasn't until the 1960s that extensive research was
conducted on the behavior and life history of these animals. This research
revealed that African elephants are social animals, organized into basic family groups of three to five members, each composed of a female and her
offspring. Several family groups together form a "clan," which may include
from six to seventy individuals, led by the alpha female or dominant sister.
Small groups of bulls may join a clan during periods of sexual reproduction,
but only by following at a distance, and they "take no part in social leadership." Akeley spent enough time observing elephants in the field to have
had an understanding of the basic family group and would have known that
grouping males and females together would not reflect the species' actual
behavior. Osborn wanted a group of four elephants, which would allow visitors to observe their sexual and age dimorphism, but more likely his real
intention was to surpass Akeley's pair of fighting bull elephants at the Field
Museum. What makes the composition compelling is that Akeley chose
to mount the family group with the male and female together at the lead.
While the old bull is alarmed by an invisible danger, the female has already
recognized it. Akeley avoided the scientifically inaccurate confines of an-

Fig. 6.1. Carl E. Akeley modeling in clay the large bull elephant for the American Museum group in 1914. (Image #34314, American Museum of Natural History Library)

other family scene for his rhinoceros groups, originally designed to flank the front and back of the elephant group, by placing the male rhinoceros on one pedestal and a female sitting or lying down with her calf on the other. The separate pedestals would visually reinforce the idea that adult male rhinos are solitary individuals, while females share a strong bond with their young and aggressively defend their calves from intruding males.[31]

In August 1914, just after Akeley began mounting specimens, the First World War broke out. Still, by 1916, Osborn could report that of the 5,800 mammal specimens collected by the Congo expedition (1909–1915) headed by mammalogist Herbert Lang and his assistant James P. Chapin, "the entire collection of skins has been permanently prepared, and some of the choicest specimens, such as the white rhinoceros and the okapi, are being superbly mounted in the atelier of Mr. Carl E. Akeley, with the cooperation of Mr. James L. Clark."[32] Lucas reported a year later, "Progress has been made on the Elephant Group, though, like all other branches of work, this has been hampered by war conditions which have called for service elsewhere those engaged upon it."[33] Akeley understood that the elephant group was an enormous undertaking, but he accepted the challenge with great enthusiasm. By year's end, he had nearly completed mounting the young bull elephant and a group of rare okapi, while Clark had completed two white rhinoceroses.[34]

Akeley would not complete the elephant group until 1921. That year, Lucas announced, "The principal achievement of the year has been the opening of the Akeley African Elephant Group, . . . on which Mr. Akeley has been engaged ever since 1909 when he left the United States for Africa to collect the materials."[35] Lucas hailed the group as "a masterpiece, both in design and in *permanence*," and observed that it gave "a surpass-

Fig. 6.2. The young bull elephant near completion in 1916. The wooden armature helps to stabilize the mount as the skin dries. (Image #36428, American Museum of Natural History Library)

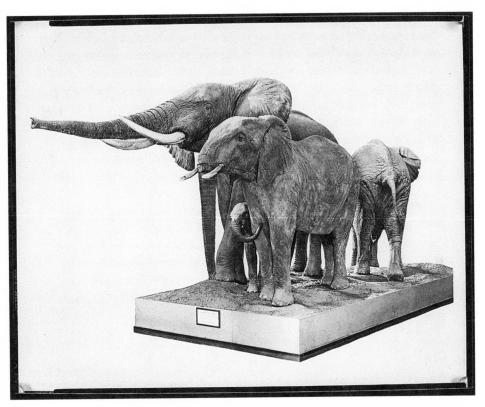

Fig. 6.3. The completed version of "The Alarm" as unveiled in 1921. (Image #310463, American Museum of Natural History Library)

ingly lifelike impression." For one year, the elephant group was shown only to potential donors in an effort to raise funds for another Akeley African expedition—this time for the express purpose of collecting the little-known, little-understood mountain gorilla. The donors responded favorably, and arrangements were made for Akeley to undertake an expedition to the Belgian Congo for the American Museum.

AKELEY'S MOUNTAIN GORILLA GROUP

Before Akeley departed on his second American Museum expedition, he met with a reporter from the *New York Times* in the museum's "elephant studio" just days before his ship sailed. The reporter found Akeley helping his assistants finish a group of lions and assessing the work remaining on the group of elephants "in one corner of the studio." Akeley "smiled enigmatically" at the elephants, now nearly complete after a decade of work. "I've studied the elephant for twenty years," he told the reporter, "and I'm not yet acquainted with him. He's a hard fellow to know. Civilization may tame him, but will never understand him."[36] On New Year's Day 1922, while Akeley was still in the Lake Kivu district of the Belgian Congo, the Akeley elephant group—featuring the world's largest land mammal and the largest group of elephants in any American museum—would be opened to the public.[37] The *New York Times* would rave:

> The artist has caught a moment in the lives of these beasts and fixed it— but not in metal or rock. That bulky trunk, so characteristically flung out, has felt for the wind many a time. Living muscles have flapped those ears. Those wrinkled skins have brushed past many an actual forest tree. . . . Lumbering, lifeless things, they fell to the ground in faraway Africa. Bundles of skin and parcels of bones, they reach the museum. Their restoration and erection as a monument to the grandeur of their living past is a story of patient and painstaking effort, covering a period of eleven years.[38]

When discussion turned toward the upcoming expedition, however, Akeley's mood changed from ruminative to palpably excited. "Our destination is entirely indefinite," he said, explaining that little was known of the Virunga Mountains, where the expedition would be hunting gorillas. "Just where we will find the best specimens or what we will find, I am unable to conjecture, as it is a new part of Africa to me—an unexplored country. In fact, this will be an entirely new adventure."[39] Akeley had obtained a license to collect ten gorillas, and he told the press that he hoped "to obtain a complete family . . . to be used in one of the lifelike habitat groups of the Museum." He also had permission "to obtain motion pictures."[40] At the very end of July 1921, the adventure began as Akeley and his crew— his secretary, Martha Akeley Miller, the big game hunter Herbert Bradley, the fiction and travel writer Mary Hastings Bradley, and the Bradleys' six-year-old daughter Alice—escaped "the torrid heat in New York" aboard the White Star liner *Baltic*, bound for Liverpool, where they would equip and continue on to the unexplored mountain regions of the Belgian Congo.[41]

The trip was a marked success. While still in the field, working from his observations, still photographs, and the first motion pictures ever obtained of gorillas in the wild, Akeley began designing the composition for the go-

Fig. 6.4. A photograph of the gorilla group as modeled in clay by Carl E. Akeley. (Image #315896, American Museum of Natural History Library)

rilla group. Mary L. Jobe Akeley, Akeley's second wife, later described the arrangement that her husband chose:

> The old male of Kirisimbi dominates the group. Disturbed by a movement in the bushes below he rises and beats his chest. The other male is shown on all fours in the normal walking attitude. One hand is poised as he hesitates in his advance. His expression is one of passive interest. One old female leans lazily against the base of a tree, while a baby idles near-by. The fifth gorilla, a second mature female, is feeding on vegetation.[42]

Compared with the stories of vicious, man-eating gorillas in the popular literature of the time, the scene was pointedly domestic.

This composition was Akeley's visual statement about the necessity of the museum taxidermist collecting his own specimens and mounting them according to his own observations. Before his trip to the Belgian Congo, Akeley believed, "the average museum" would have purchased gorilla skins from big game hunters, and the preparators "would have studied the available writings on gorillas. They would have found out that the gorilla was a ferocious animal who inhabited the dense forests and, like as not, that he lived in trees most of the time. And that is the kind of animal the group would have shown."[43] By contrast, in mounting his gorilla group, Akeley would rely only on his own direct observation of the species in its environment:

> My own measurements are significant and helpful. I have photographs of the scenery, the setting, and the gorillas themselves. I have photographs of their faces—not distorted to make them hideous but as they naturally were—and death masks which make a record that enables me to make the face of each gorilla mounted a portrait of an individual.[44]

By matching each specimen to its field measurements, mounting the skin of each face over the death mask of that particular individual, and posing the specimens in attitudes drawn from film and still photographs, Akeley sought to create a "true and faithful copy of nature."[45] Perhaps he was hoping to achieve what his predecessors had not: scientifically accurate mounts that could serve a dual purpose as both exhibits and study specimens.

Some contemporary critics disparaged the chest-beating male as sensational. However, the depiction was—and has continued to be—misinterpreted. Akeley described witnessing chest-beating behavior on the 1921 expedition and "making a motion picture record of it." In the film, a female with two of her young rises momentarily from her perch to beat her chest, then returns to "making herself comfortable with the apparent intention of going to sleep." Akeley concluded that the behavior was merely "a nervous expression of curiosity."[46] Mary Akeley later echoed this view, writing that chest beating "seems to denote curiosity or to serve as a warning to the other members of the family." She had seen the same chest-beating behavior during a later expedition and asserted that "it has never been coupled with any aggressive act."[47] To her husband's critics, Mary replied that the gorillas could have been "much more spectacularly mounted" and "much more startling in their appeal" had Akeley been "inclined to accept the traditional view of the gorilla." Instead, Akeley had meticulously represented every aspect of the gorillas' anatomy, behavior, and environment, because "in his eyes it was nothing short of a crime to place in an educational institution like the American Museum of Natural History an exhibit that lacked a basis in natural history fact."[48]

But Akeley was hardly averse to creating controversy. Back at the museum, while he and his assistants were engaged in mounting the specimens for his gorilla group, he was also at work on a bronze sculpture he titled "The Chrysalis." The statue depicts the figure of a man—a close likeness of Akeley himself—emerging from the skin of a mountain gorilla. Akeley was probably drawn into the country's debate over evolution—which would culminate in the famous Scopes Monkey Trial in July 1925—particularly when William Jennings Bryan and Henry Fairfield Osborn, the American Museum president, entered into an extraordinary public debate in the Sunday editorial section of the *New York Times*.[49] Bryan excoriated evolutionists for their adherence to Darwinism, or "evolution applied to man." His main objection to Darwinism was that it would make agnostics out of Christians, just as it had made an agnostic out of Darwin. Bryan lamented, "He brought man down to the brute level and then judged man's mind by brute standards."[50] Akeley chose not to address Bryan directly as Osborn did. Instead, he went to work in the studio and submitted "The Chrysalis," as he had many of his other bronze sculptures, to the National Academy of Design. But the Academy refused to display the sculpture because of its "evolutionary content." In response, Akeley's friend Reverend Charles Francis Potter, of the West Side Unitarian Church, declared Sunday, April 27, 1924, "Evolution Day." As part of the daylong event, "The Chrysalis" would be publicly unveiled, and Akeley would deliver a lecture titled "Personality in Animals."[51]

Akeley's topic was carefully chosen. When Dr. John Roach Straton of Calvary Baptist Church first learned of Potter's plans, he had denounced the celebration as a "glorification of bestiality."[52] Potter and Straton, beginning in December 1923, had engaged in their own series of debates regarding evolution after Straton had charged the American Museum with "mis-spending the taxpayers' money, and poisoning the minds of school children by false and bestial theories of evolution." Osborn was a staunch opponent of Straton and Bryan, but he was "not willing to have Mr. Akeley's 'The Chrysalis' connected with the name of the American Museum of Natural History" because the museum did not have "the authority to express opinions on works of art."[53]

Akeley was undeterred by Osborn's lack of support. Because Straton had likened Akeley's statue to the "bestial" work of Alyce Cunningham—a British woman who raised gorillas in her home, observing and testing their mental abilities—Akeley seized on the opportunity to meet with her after

Fig. 6.5. "The Chrysalis," sculpted by Carl E. Akeley. (Image #249306, American Museum of Natural History Library)

William T. Hornaday at the Bronx Zoo invited Cunningham and her companion, the three-year-old gorilla "John Sultan," to visit New York. Upon their arrival, Akeley greeted Cunningham at the docks and helped her deliver John Sultan to his own suite at the Hotel McAlpin. Recognizing an opportunity to promote his own work, Akeley notified the press and introduced them to Cunningham and the young gorilla. "There is the living thing that is nearer to man than anything else," he told the assembled reporters. "Let those who do not believe the Darwin theory of evolution look at that animal, and then doubt that he and man had at some time a common ancestor."[54]

The next week, Akeley convinced Osborn to allow Cunningham and John Sultan to come to the American Museum for a press conference. Akeley assembled the reporters in his taxidermy studio, where two of the five specimens for his gorilla group had already been completed. "These giants were mounted with marvelous naturalness," the reporter for the *New York Times* wrote. "But little John showed no interest in them. They smelled of arsenic and other chemicals used in preserving the skin and driving away moths."[55] Nor did he show any recognition of the death masks of the five gorillas cast in plaster. Akeley had hoped for a photograph of himself with the young gorilla and the two completed mounts, but John Sultan would not sit still long enough for any of the six photographers to capture the moment. Nevertheless, Akeley had seized the opportunity to forward his belief in evolution and demonstrate the docility of gorillas.

With the flurry of publicity, more than six hundred people attended the unveiling of "The Chrysalis" on "Evolution Day." When Akeley addressed the crowd, he began by explaining his intention with his sculpture. "I do not mean to suggest that man sprang from the gorilla," he said. "They, undoubtedly, had a common ancestor. Science is on the trail of this ancestor and will locate it." He then denounced Straton's assertions that gorillas were "bestial." He declared, "You do not find bestiality among animals; only among human beings. If people knew animals as I do, they would never misuse the word bestial." He concluded by relating a painful episode from his gorilla-hunting expedition. He had shot a female, and one of his native assistants had speared her young. "I came to it as it lay dying," Akeley told the hushed audience. "Its mother was already dead. When it saw me it stretched out its infant arms in appeal and cried when I touched it. Is that your 'bestiality' or is man the beast?"[56]

That infant was the last gorilla Akeley ever killed. Though he had been permitted to collect ten specimens in 1921, he took only five. However, Akeley's frequent depictions of the timid, easily hunted gorillas had had an unintended effect: among white game hunters, the certainty of obtaining such a rare trophy had touched off a surge in gorilla hunting. Since 1922, Akeley had been lobbying the Belgian government to outlaw the commercial hunting of gorillas and set aside a sanctuary for them. With the publicity surrounding "The Chrysalis," Akeley began making that plea publicly. In June 1924, the *New York Times* ran an editorial quoting Akeley and denouncing the hunting of gorillas. "No beast is so well worth studying," the editor wrote. "It is in danger of extinction. Belgium is largely responsible if the last go- rilla in the Congo country has been killed."⁵⁷ In September, Belgium yielded to the international pressure and informed Akeley that it had approved his plan to establish a 250-square-mile gorilla sanctuary around Mount Mikeno, named Parc National Albert for King Albert I, who had approved the pro- posal. The plan had been endorsed not only by the Belgian ambassador, but also by William T. Hornaday, who had offered to build facilities at the Bronx Zoo to accommodate gorillas and organize international scientists in their study. T. Alexander Barns at the British Museum proposed to attempt to cap- ture a mate for John Sultan in an effort to begin a captive breeding program.

On March 2, 1925, Parc National Albert was officially established by royal decree. Akeley praised Belgium's decision in a special issue of the travel magazine *The Mentor*, dedicated to his achievement:

> For years the gorilla has been protected from man's attack by an unearned reputation for ferocity. . . . Deprived of this protective disrepute as a re- sult of my observations and experiences in the Kivu the gorilla was left defenseless. No longer considered invincible he became merely another game animal, the more eagerly hunted down because of the novelty of the experience. Had not the Belgian Government realized the serious- ness of the situation I believe the Kivu gorillas would have disappeared.⁵⁸

On the heels of this great success, Akeley was eager to return to the Bel- gian Congo to collect the foreground accessories he needed to complete his gorilla group for the American Museum's African hall. He quickly found fi- nancial backing from George Eastman and Daniel E. Pomeroy to lead a spe- cial expedition that would collect not only accoutrements for the gorilla group, but also the necessary specimens for six new groups for the African hall.⁵⁹

While making arrangements for this expedition, Akeley completed his lion group and installed it in the temporary African hall "as a model for the type of groups that are to constitute the great African Hall."⁶⁰ The *New York Times* described the group with wonderment:

> The cubs stretch their necks eagerly at a couple of small water holes in a field of stubble and waving grass. Behind them dawn breaks across the deep blue sky. This is a picture, complete with background and fore- ground and central figures. Against the painted sky is grass that actually grew in Africa. The animals lived and died there. The tableau has color, life, vitality—everything except motion and sound.⁶¹

Fig. 6.6. The gorilla group in the Akeley Memorial Hall of African Mammals. Completed in 1936, is still on display at the American Museum of Natural History. (Image #6918, American Museum of Natural History Library)

Three of Akeley's bronze lion-hunter sculptures were moved into the same space, along with the elephant group, the white rhinoceros group, and a model of the future African hall. As a final touch, the five mounted gorillas were assembled in a temporary case, which Lucas reported "made it practicable to set aside the North Asiatic Hall as a Pro-African Hall, which was formally opened on January 21, 1926."[62]

Soon after, Akeley left for Africa—never to return. He would not see his African hall, nor his beloved gorilla group, completed. Nevertheless, the hall's design was entirely Akeley's—and was intended to counter the image of the continent conveyed in the nineteenth-century adventure writings of Paul Du Chaillu and in Henry Morton Stanley's *In Darkest Africa*. Akeley condemned their illusory language about Africa and especially its gorillas, which they portrayed as aggressive, and it was with an overt disdain of Stanley's imperialist view of the African continent that Akeley wrote *In Brightest Africa*. In the chapter "Is the Gorilla Almost Man?" he deconstructed a passage from Du Chaillu's *Wild Life Under the Equator* in order to disarm the popular view that gorillas were aggressive and dangerous.[63]

Akeley effectively quotes the passage by placing in brackets what obviously represents Du Chaillu's own interpretation or "feelings" about the species, and leaving outside the brackets only "what the gorilla did":

> [And now truly he reminded me of nothing but some hellish dream-creature—a being of that hideous order, half man, half beast, which we find pictured by old artists in some representations of the infernal regions.] He advanced a few steps—then stopped to utter that [hideous] roar again—advanced again, and finally stopped.[64]

Akeley interpreted "what the gorilla did" as scientific fact, and through his own scientific observations, he hoped to sort out fact from fiction.

For Akeley, it was important to observe the gorillas' true behavior, as the species was "unquestionably the nearest akin to man," and he anticipated that the specimens he collected would provide "important opportunities" for research in comparative anatomy, psychology, medicine, and natural history.[65] It was his goal to unravel myths about the species that had been almost a century in the making, to study the mountain gorilla in its native habitat, and to bring back a small number of specimens for scientific research, as well as camera footage.

In fact, by 1923, Akeley was advocating a new way of "collecting" wildlife for scientific research. "Camera hunters appeal to me as being so much more useful than the gun hunters," Akeley wrote in *In Brightest Africa*. "According to any true conception of sport—the use of skill, daring, and endurance in overcoming difficulties—camera hunting takes twice the man that gun hunting takes."[66] Clearly Akeley had fallen in love with Africa and its wildlife, and like so many naturalists of the period, he feared that overhunting was leading to extinction, especially of the large mammalian species. The new technology of moving pictures, he believed, not the old technology of the hunter's gun, offered a possible alternative to collecting specimens for scientific research.

THE AKELEY-EASTMAN-POMEROY EXPEDITION

In 1923, *Trailing African Wild Animals*, filmed by famed adventurers and commercial cinematographers Martin and Osa Johnson, first debuted at America's premier movie house, Capitol Theatre, on Broadway in New York City. Martin Johnson, whose camera of choice was the Akeley Motion Picture Camera, had met the inventor two years earlier at a gathering of the Explorers Club. Concerned about the fate of African wildlife, Akeley had urged

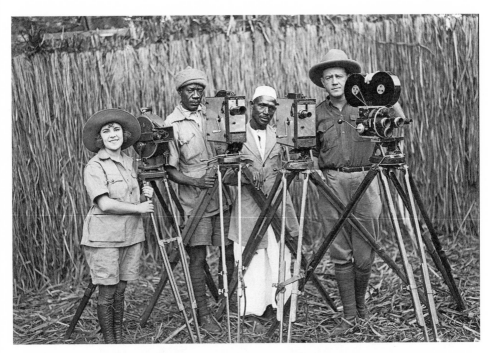

Fig. 6.7. Martin and Osa Johnson and African attendants with field cameras in Kenya, 1923. (Image #129104, American Museum of Natural History Library)

Johnson to employ his considerable skills as a cinematographer to make a record of African animals in their natural habitats. The two formed an alliance that would culminate in Johnson's film and Akeley's fifth African expedition. Believing that motion pictures would help promote Africa and, ultimately, the American Museum's African hall, Akeley convinced Henry Fairfield Osborn to endorse *Trailing African Wild Animals* and Martin Johnson himself as a serious wildlife cinematographer with "no superior." The endorsement helped bring the film—billed as "The World's Most Perilous Camera Expedition," and featuring Osa Johnson as the heroine—to a wider audience. With its support from the scientific community, the film also captured the attention of many of New York's wealthiest entrepreneurs, including the septuagenarian George Eastman of Eastman Kodak Company in Rochester, New York. When Johnson approached Eastman to help fund the making of a second African film, he agreed to the sum of $10,000.[67]

Capitalizing on Eastman's interest, in early summer 1925, Daniel E. Pomeroy, an AMNH trustee and the director and vice president of Bankers Trust Company, approached Akeley with an invitation to organize and guide an African safari funded by himself and Eastman, who now wished to see Africa firsthand, guided by Akeley and accompanied by the Johnsons. Akeley had received many such requests from trustees in the past, but had always declined. This time, however, he agreed, on condition that Eastman and Pomeroy also finance the collection of specimens and accessories for the museum's African hall. The funds were to cover the collection of specimens, "studies and data . . . for background and accessory material," and the "mounting and installation of the groups complete in the Museum." Pomeroy and Akeley traveled by rail to Rochester to discuss Akeley's plan with Eastman. When it came time to ask Eastman for funding, Akeley asked for the full sum needed for the African hall—one million dollars. Eastman, however, guaranteed only the initial sum of $100,000 to finance two of the four large corner groups and one smaller group. Pomeroy agreed to fund a group of greater kudu. Akeley was delighted with the initial promise and hopeful that Eastman, like himself, would fall in love with Africa's beauty and eventually be willing to fund the entire completion of the African hall.

Although Akeley had been on four previous African expeditions, he intended this one to be unlike all the others. Just as the taxidermist must view the animal in the wild, Akeley believed that the accessories preparator must view the flora and geology in the field, and the background painter must see the natural landscape. "The background—and it is a beautiful scene—must be painted by as great an artist as we can get and he must go to Karisimbi to make his studies. And the preparators who make the accessories—the artificial leaves, trees, and grasses—they, too, must go to examine the spot and collect their data."[68]

Akeley also required a well-trained preparator-taxidermist to assist him in the field. He set out to assemble a premier team—one that would work well together, each with "energy, common sense, a special ability, and great love for the duties at hand."[69] For Akeley, the expedition was also exceptional because it would finally make it possible to realize his vision and demonstrate his innovative taxidermy techniques, "For the first time we have the opportunity to train a group of men not only to practice the various arts which are combined in making modern zoological exhibits, but also to further develop the methods that make this sort of museum exhibition worthwhile from the scientific and artistic standpoint."[70]

As luck would have it, while Akeley was making arrangements for his trip, he was called on by taxidermist Robert H. Rockwell—who had trained in Akeley's methods under the Santens brothers at Ward's, worked with Ward's alumnus Nelson R. Wood at the National Museum, and served as chief taxidermist under Lucas at the Brooklyn Museum. In short, Rockwell's

Fig. 6.8. Carl E. Akeley and Mary Jobe Akeley at the American Museum of Natural History, 1926. (Image #311321, American Museum of Natural History Library)

experience and skill made him just the kind of preparator Akeley was seeking for his growing corps of highly trained exhibition professionals. Rockwell later remembered that when he told Akeley he was looking for work, "he surprised me by revealing that I could have a job right away if I wanted it."[71]

In January 1926, Akeley and Mary Jobe Akeley traveled ahead of the main party, "attending to preliminary details relating to the expedition in

London, thence to Nairobi, to complete arrangements for the safari of the main expedition."[72] Rockwell sailed for London in March, gathered miscellaneous material requested by Akeley, and then sailed aboard a steamer of the Union-Castle Line for British East Africa. Eastman and Pomeroy joined the steamer in Genoa, and together the three men traveled to Mombasa, where Akeley was waiting to take them by train to Nairobi, where they would rendezvous with Mary, Martin and Osa Johnson, and their outfitters, the renowned safari hunters A. F. "Pat" Ayre and Philip Percival.[73] It was May by the time they moved on to Seronera in the Great Rift Valley, where Eastman and Pomeroy would establish camp and hunt for "trophies" with Ayre and Percival, while Akeley and Rockwell moved on to the first "museum camp" at Lukenya Hill, an inselberg on the plains southeast of Nairobi, where William R. Leigh, a famed painter of the American West, and his colleague Arthur A. Jansson were already at work.[74]

Akeley's general blueprint for designing habitat groups guided the activity of the bustling museum camp. He had established that the site depicted in each wildlife group should be "typical" of the animal's habitat, with consideration given to "practicality of reproduction of foreground accessories (vegetation)."[75] He did not wish to reproduce idealized settings, but rather to show an exact reproduction of a given locality. Plaster casts of the vegetation (usually individual blades of grass, leaves, and flowers) were made in the field and transported back to the museum, where they were used to create wax reproductions.[76] The painted background, he believed, should depict the "character of landscape and wherever possible showing something of historic or geographic interest."[77] With Akeley's guidance, Rockwell selected vegetation and made "plaster casts . . . for wax reproductions to be used in the foregrounds of the groups," and Leigh and Jansson painted "striking views" and panoramas of the African landscape.[78]

Akeley's first project was designing the klipspringer group, the first of the smaller groups. He made a "sketch model," which included "background, models of the animals, vegetation—a complete study in miniature."[79] Akeley intended the model to be used by the exhibit preparators; once they had returned to the museum, they would only have to "reproduce this sketch in full size."[80] Akeley also planned whenever possible to film the very specimens intended for the group in their natural habitat to guide the taxidermist in mounting each specimen. He succeeded at this location in filming the four klipspringers used in the museum group.

Akeley had anticipated collecting reedbuck at this camp, but none were found. His worst fears had been realized: the once plentiful populations of large plains animals had declined significantly in the twenty years since his first expedition. He lamented that white settlements were encroaching on the once remote region, decreasing habitat, while overhunting by commercial white hunters was further depleting large mammal populations. As Rockwell recalled, "Literally thousands of zebras were slaughtered annually for their hides. And many more were ruthlessly destroyed because of the allowance in those times of twenty head on each full hunting license."[81] In September 1926, Akeley had a startling realization—one that William T. Hornaday had experienced in the American West forty years earlier:

> I have just come in from a two days' trip down the Tana, through a region I have known only as swarming with game, but I now find it a complete waste. There is only a pitiful remnant of the great buffalo herds of the past, and of the other game almost nothing. This is a condition we have found everywhere we have been in Kenya Colony. I have not appreciated the absolute necessity of carrying on the African Hall, if it is ever to be done, as I now do after this painful revelation.[82]

With the sense that it was more important now than ever to complete the African hall, and with the knowledge that Eastman would probably not provide any further funding, Akeley stepped up his efforts, pushing himself to his physical limits.

In June, the party moved on to a location on the Euaso Nyiro River, intended as the site of the large water-hole corner group, which would include both Grevy's and Grant's or plains zebras, "Grant's gazelle, a dik-dik, and three tall reticulated giraffes."[83] The process of designing, modeling, filming, and collecting for the habitat group began again. While hunting with Eastman on the Serengeti Plain, Akeley fell ill with fever, which was later diagnosed as the result of a "nervous breakdown." He was transported by his wife Mary to the hospital in Nairobi. After finding no vacant beds there, she was forced to take him to the Kenya Nursing Home, where he remained for three weeks.[84]

While Akeley continued to recuperate, Rockwell and Pomeroy went on to Kidong to collect specimens of the African buffalo and kudu. After another two weeks passed, Akeley, feeling rested, decided to reschedule the expedition. Before leaving the city, Akeley and Mary rendezvoused with Belgian zoologist Jean Marie Derscheid to make plans for the Congo segment of the expedition. In spite of his prolonged illness and weakened condition, Akeley insisted that they return to the Virungas.

The group departed for the Congo on October 14, 1926. The month-long trek along old caravan routes was arduous, and was made even more difficult by periodic rains. But the Virungas and the gorilla sanctuary called to Akeley. By the time they reached the lower slopes of Mount Mikeno in early September, the cold rains had weakened Akeley, and his fever had returned. Mary, too, had fallen ill. As the cold, damp mist hung low and the red glow of the Virungas' active volcanoes penetrated it, the trail must have taken on a surreal quality for the Akeleys. But still they pressed on. Near Kabara, Akeley sent for Leigh, who would paint the panorama for the gorilla habitat group at the site where, years earlier, Akeley had collected the "old male of Kirisimbi." As they neared the summit, Akeley had difficulty maintaining his footing on the steep trail, so his companions carried him in a hammock for a time, until, drenched by rain and chilled to the bone, Akeley insisted on walking the last few miles. As they trekked on, Akeley and the group startled a family of gorillas feeding on wild celery. Akeley was thrilled to share the sighting with Mary. Upon reaching the saddle between Karisimbi and Mikeno, they made camp and rested.

As Akeley lay on his cot, delirious, racked with violent dysentery, and hemorrhaging blood, he spoke of the museum and its electrical projects. The following night, as the snowy peak of Karisimbi shimmered in the moonlight outside his tent, Akeley fell unconscious, and his heartbeat slowed. By the time Leigh arrived, Akeley was dead. The American Museum received Mary Jobe Akeley's telegram from the Belgian Congo on November 30: "My husband's spirit passed Nov. 17; hemorrhage; Slope Mikeno. I remain supervise completion of background accessories gorilla, koodoo, according his plan. Inform family, friends."[85] Though devastated and in shock, Mary and the expedition members buried Akeley there in the saddle between the two mountains, in the heart of the Virungas, where he had once said, on the 1921 expedition, that he hoped he would one day die. William Leigh's spectacular painting—now the background for the gorilla diorama in the Akeley Hall of African Mammals—memorializes the site where the "old male of Kirisimbi" once lived, and where Carl Akeley died.[86]

CONCLUSION

Despite Akeley's untimely death, he had so thoroughly planned every aspect of the exhibit process that the work could be carried on without him. Years later, Rockwell recalled, "Ours was the only expedition that I know of that brought back its entire collection with cased skins and a complete skeleton of each specimen for guidance in the final mounting."[87] Even the death of Frederic Lucas in 1929 could not derail or delay the project. By that point, the commitment of the institution to this new method of museum display was complete.

In fact, Akeley's demise seems only to have steeled the resolve of his protégés to see their fallen mentor's vision for the African hall realized. Painter William Leigh's fervor and rhetoric, for example, began to sound strikingly like Akeley's own. He told readers of the American Museum's journal *Natural History*:

> Not only must the backgrounds be correct, but they must be as typical of the continent as were the beasts they accompanied; in fauna and flora, in geology and geography, we must give as comprehensive a sense of the essence of Africa as possible within our limitations. We must produce complete pictures, faultless history, perfect science.[88]

His goal, like Akeley's, was to create a hall of enduring permanence, a monument that would "survive, perhaps after much of this animal life has been wiped out."[89] Akeley's vision of a corps of well-trained and committed preparators working together to produce the finest museum habitat groups in the world would soon be realized.

For the next eleven years, Rockwell, now under James Clark's direction, "worked continuously, modeling and mounting" African specimens— "thirteen groups in all, or about half the animals in the African Hall, including four elephants."[90] The Akeley Hall of African Mammals finally opened in 1936, but the hall's impact was much more far-reaching.

Even before its completion, Akeley Hall became the template for creating new African exhibits in all American natural history museums. As Karen Wonders has noted, the rapid disappearance of African wildlife and the race to enshrine it in the halls of natural history museums "captured the imagination of the time."[91] John Rowley, chief of exhibits at the Los Angeles County Museum of Natural History, for example, acknowledged Akeley's work as the inspiration for a hall of African wildlife that he began in 1920 and finally completed in 1928. After Rowley's own untimely death, his main benefactor, Leslie Simson, convinced the California Academy of Sciences in San Francisco to undertake a similar hall—opened in 1934. In the meantime, the Field Museum of Natural History used Akeley's mounts to create its own African hall, opened in 1932 and named Carl E. Akeley Memorial Hall.

However, the influence of Akeley's diorama idea was hardly limited to African exhibitions. "Even within the AMNH itself," Wonders writes, "the impact that Akeley's grand hall had on exhibition was considerable."[92]

The Hall of South Asian Mammals, in particular, designed and executed by Clark and Rockwell, was meant to serve as a direct companion to the African hall, but within five years of the completion of Akeley Hall, the AMNH had also added the Birds of the World Hall, the Whitney Memorial Hall of Oceanic Birds, and the Hall of North American Mammals—all modeled after Akeley's design.

By the middle of the twentieth century, the diorama method—with its accurately modeled specimens, realistic foliage and foreground materials, arched and painted backdrop, and interpretive label—was the dominant method of display in the American natural history museum. Even after the development of high-quality color photography and the prevalence of motion pictures brought the world's wildlife into our homes through magazines such as *National Geographic* and popular television programs such as *Wild Kingdom*, these striking and dramatic dioramas have continued to draw generations of visitors. To this day, more than three million people visit the Akeley Hall of African Mammals each year, and the museums with exhibits modeled after Akeley's dioramas—in all parts of the United States and around the world—attract many millions more.[93]

To my Illustrious Successor: . . . When I am dust and ashes I beg you to protect these specimens from deterioration & destruction. Of course they are crude productions in comparison with what you produce, but you must remember that at this time (A.D. 1888, March 7) the American School of Taxidermy has only just been recognized. Therefore give the devil his due, and revile not.

—William T. Hornaday[1]

More than a century has passed since the taxidermists at Ward's ushered in a new era for animal display in American natural history museums, yet we continue to feel their impact. To this day, it is impossible to visit America's major metropolitan natural history museums without seeing their work—and the work of those they trained—on prominent display. The American Museum is still acclaimed for its halls of African, Asian, and North American wildlife, designed by Carl E. Akeley, James L. Clark, and Robert H. Rockwell; the renovated group of six elephant seals collected by Charles H. Townsend and mounted under his direction can still be found in the Hall of Ocean Life. At the Field Museum, Akeley's fighting bull elephants dominate the main hall, and most of his habitat groups remain on exhibit, including "The Four Seasons." The Carnegie Museum continues to feature many of the specimens mounted by Remi and Joseph Santens, as well as Frederic S. Webster's "California Condors and Turkey Buzzards on Dead Wapiti" and Jules Verreaux's "Arab Courier Attacked by Lions," as restored by Webster. The Milwaukee Public Museum removed Webster's original flamingo group, but several of Akeley's displays—including the muskrat group—remain on exhibition. Hornaday's "A Fight in the Tree-Tops, along with many other Akeley-style dioramas," was exhibited at the

National Museum of Natural History until 2000, when the museum began installing its new Kenneth E. Behring Family Hall of Mammals. Many smaller institutions have benefited indirectly from the work of these taxidermists as well; as the large metropolitan museums for which they worked have deaccessioned some of their mounts, they have been donated or loaned to such diverse institutions as the University of Iowa Museum of Natural History, the National Bird Dog Museum in Grand Junction, Tennessee, and the Museum of the Northern Great Plains in Fort Benton, Montana.

In the brief history of American museum taxidermy, the natural history museum shifted its focus from pure research to include dissemination of knowledge through exhibits and public education. Ward's taxidermists—William T. Hornaday, Frederic A. Lucas, Charles H. Townsend, Frederic S. Webster, and Carl E. Akeley—shaped and defined the public side of American natural history museums, zoos, and aquaria through their technical advances in taxidermy, innovative exhibit design, and distinctive educational content. As they transformed the work of taxidermy from a trade to a museum profession, they revolutionized methods of animal display, transforming exhibits from rows of stuffed specimens with scientific labels to lifelike mounts arranged in family groups, supplemented with photographs and descriptive labels. At the same time, they conceived of an educational directive that would not only teach museum visitors about the natural world, but would instill in them an appreciation for the human role in nature—specifically, the responsibility of humans in preventing the extinction of species.

Since Hornaday's 1885 awakening to the impending extinction of the American bison, numerous species have been pulled back from the brink of extinction, while still others around the globe have lost the struggle. From 1875 to 1925, forty-seven species are known to have gone extinct worldwide, including the Labrador duck, the Falkland Islands dog, the red gazelle, the Guadalupe caracara, the passenger pigeon, and the Carolina parakeet.[2]

For some sense of the toll extinction might have taken in North America, one need only look at the example of Australia. During the period from the 1880s to the 1930s—the same period during which Hornaday, Lucas, and Townsend were actively advocating for the protection of North American wildlife—the fauna of Australia, without any protective legislation, was devastated. This brief period saw the extinction of the Eastern hare-wallaby (1889), the short-tailed hopping-mouse (1896), the pig-footed bandicoot (1901), the long-tailed hopping-mouse (1901), the robust white-eye (1918), the Paradise parrot (1927), the lesser stick-nest rat (1933), the desert rat-kangaroo (1935), the thylacine (1936), and the Toolache wallaby (1939).[3]

The North American species saved from extinction as a result of the pioneering conservation work of Hornaday, Lucas, Townsend, and Webster include the American bison, the West Indian monk seal, the northern elephant seal, the northern fur seal, the blue whale, the right whale, the snowy egret, the whooping crane, subspecies of the sandhill crane, the brown pelican, and the American flamingo. In Africa, if not for Carl Akeley's foresight in urging the Belgian government to protect critical mountain gorilla habitat in the Virunga Mountains of the Belgian Congo, the mountain gorilla would surely have gone extinct in the first half of the twentieth century. Parc National Albert, today Virunga National Park in the Democratic Republic of the Congo, is one of only two national parks protecting the world's population of about 700 mountain gorillas. The 2018 mountain gorilla census estimated a total of 604 individuals living in greater Virunga, demonstrating the critical role that the park plays in protecting the species, as it continues to teeter on the brink of extinction—constantly threatened by political instability and war.

While historians of conservation largely credit Hornaday with saving the American bison from extinction, and Akeley with preserving critical gorilla habitat in the Virungas, Lucas and Townsend have been overlooked almost entirely, perhaps because they participated in the American wildlife conservation movement for the most part from behind the political front lines. Nevertheless, their scientific research saved many endangered marine species. Their combined contributions to understanding the natural history of the fur seal were critical to the success of the 1911 fur seal treaty that banned pelagic sealing—the first international treaty to conserve wildlife, which set a positive precedent for international cooperation on future conservation issues. In addition, Townsend's tireless work on behalf of the northern elephant seal finally won its protection when Mexico passed legislation in 1922 to protect the species. Like Hornaday, Lucas and Townsend also contributed to the public's understanding of the severe impact humans could have on wildlife. Lucas published popular books, such as *Animals of the Past* (1901) and *Animals Before Man in North America* (1902), that helped answer questions about extinction for a wide general audience, while Townsend presented public lantern slide lectures about the world's oceans and endangered marine mammals.

Similarly, Akeley's popular account of collecting in Africa, *In Brightest Africa* (1923)—a compendium of ten years of his publications in *The Mentor* (a magazine similar to *National Geographic*), and *American Museum Journal* (the magazine of the AMNH), among others—helped to dispel negative gorilla myths and establish the need for an international movement to conserve mountain gorilla habitat. Although Akeley died before completing his work, his wife, Mary L. Jobe Akeley, continued to study mountain gorillas, and later published *Carl Akeley's Africa: The Account of the Akeley-Eastman-Pomeroy African Hall Expedition of the American Museum of Natural History* (1929) and numerous other books urging the conservation of Africa's wildlife.

The last twenty years have seen millions of dollars raised and many projects undertaken and completed to restore and preserve the work of former Ward's taxidermists at the U.S. National Museum of Natural History, the Carnegie Museum, and the American Museum. In 1995, the Carnegie Museum opened its new Hall of North American Wildlife on the museum's second floor, which replaced its original mammal hall designed by Remi Santens ninety years earlier. The new hall emphasizes biodiversity and the need for conservation of mammalian fauna in North America. Habitat dioramas continue as the dominant design. The main attraction is a renovated version of Remi Santens's Alaskan brown bear diorama.

The exhibit's new design was based on the field research of an exhibit designer and a plant preparator, who traveled together to Kodiak Island to observe the bears in their natural habitat—just as Akeley's design team traveled with him to Africa in 1926. The two also took photographs and collected new accessory material to replace the originals, which had since faded irreparably. A new painted panoramic background was also commissioned.[4] Additional specimens of the various animals that naturally inhabit this region—including a kingfisher, a short-tailed weasel, and glaucous-winged gulls—were collected and mounted to show the diversity of animal species in a single habitat. The composition of Santens's family group was altered to reflect today's scientific knowledge of bear behavior. With the male now posed outside of the exhibit on a rock outcropping, the museum visitor is encouraged to walk on a simulated sandbar, between the male and the female with cubs, where the rocks of the creek bed inside the diorama reach beyond the glass. The scene is made complete with sounds of roiling creek water and screeching gulls in the distance. To address educational subjects not represented in the dioramas, such as animal classification and mating, exhibit designers added an "educational area" with videos and interactive units.[5]

In 2002, the AMNH received a large Getty Foundation grant to undertake a conservation survey of the dioramas in the Akeley Hall of African Wildlife. Steve Quinn, museum artist and senior project manager for the AMNH Department of Exhibition, managed a team of conservators charged with identifying conservation issues in individual taxidermy mounts, wall paintings, and foreground accessories. The elephant group exhibited out in the open at the center of the hall was found to be in poor condition compared with the glass-enclosed specimens in the surrounding dioramas. Nearly a century of exposure in an uncontrolled ambient environment had resulted in cracked tusks, splitting and gaps in the skin, and loss of hair, as well as a thick layer of dust over the eight mounted specimens.[6] In contrast, the mounts, wall paintings, and accessories within the dioramas were found to be in relatively good condition. However, conservators found that lighting in the sealed diorama environment posed unique threats, particularly to the specimens, as it led to high heat loads and fluctuations in relative humidity. State-of-the-art cleaning methods and modifications to the lighting and HVAC systems helped to stabilize the hall. Unlike that of the Carnegie, the AMNH administration announced that it intended to preserve the original design envisioned by Carl Akeley.[7]

While administrators for the CMNH and AMNH chose to preserve their now historic dioramas in perpetuity, the directors of the National Museum of Natural History elected to undertake a complete redesign of the museum's Hall of Mammals, which included entirely new taxidermy mounts. When the curtain lifted to reveal the new Kenneth E. Behring Family Hall of Mammals in 2004, visitors saw a hall devoid of painted dioramas. For the first time in nearly seventy years, the habitat diorama, so familiar to museum visitors, was not the central theme of a major American natural history museum's wildlife exhibit. Instead, the exhibit designers chose to display mounted specimens in a newly restored hall, open to its skylights and accentuating its original Beaux-Arts design. Rather than emphasizing family groups and how they relate to habitats, the new design featured biodiversity and the similarity of certain species across habitats. Despite the shift in presentation, lifelike taxidermy mounts are still the exhibit's main attraction.[8]

Akeley's dream of a group of naturalist-artists working together to create a single unified design has been realized again and again throughout the twentieth and now twenty-first centuries. Across the United States, natural history museums—public, private, and university—have created permanent exhibit departments with public program directors to design rotating exhibits and keep permanent exhibits up-to-date. The redesign of the National Museum's Hall of Mammals involved a team of over three hundred individuals, including five taxidermists, exhibit designers, and biologists. The National Museum's two full-time taxidermists mounted over two hundred specimens. As in the past, the hall's donor, Kenneth E. Behring, collected many of the mammal skins on safari. Others were obtained from the museum's own exhibit collections and from the Smithsonian's National Zoo, while the remainder came from museums worldwide. To make room for the new hall, all of the habitat groups were dismantled. The specimens were either destroyed or removed to the immense Museum Support Center in Suitland, Maryland. Hornaday's group "A Fight in the Tree-Tops" now resides at Suitland, dismantled and crated in climate-controlled storage. The taxidermists did restore about twenty percent of the previously exhibited taxidermy mounts, including one of Roosevelt's white rhinoceroses, collected in 1909, which had been mounted by Carl Akeley's protégé James Clark.

Today's taxidermy techniques have not significantly changed from those of the "new taxidermy" developed by Hornaday and Akeley, but new materials, such as plastic and foam, have replaced the metal and wood armatures, simplifying the process. Plaster and clay are still used, especially in re-creating facial features, which is much like the technique of three-dimensional facial reconstruction employed by forensic anthropologists. Death masks (first used by Akeley), body measurements from the field, and photographs are still used to mount anatomically accurate specimens. In many ways, however, taxidermy is rapidly becoming a lost art. After the completion of Behring Family Hall of Mammals, the National Museum downsized its taxidermists, and none of the other major natural history museums in America employ full-time taxidermists anymore. Most institutions now feel that accurate depictions of wildlife and its habitat are the purview of photography and videography, and that taxidermy—expensive, time-consuming, and difficult to complete with artistic skill—is no longer central to their mission.

Almost a century has passed since Webster and Lucas debated the use of photographs as supplementary instructional material in natural history museum exhibits and since Akeley invented the first motion picture field camera. Akeley's camera, which marked the beginning of wildlife cinematography, changed the museum audience from passive visitors, who came with misconceptions about the natural world based on popular myths, into active observers with a more comprehensive knowledge of nature. By the mid-twentieth century, natural history museum administrators began to focus educational directives toward a younger, less experienced audience. By the 1970s, the new field of museum education had emerged. Professional educators began to develop innovative methods to engage the museum visitor in a more dynamic learning environment. Exhibits began to feature instructive materials designed to engage as many of the human senses as possible, including photographs, digital video, sound, and interactive hands-on experiences.[9]

Today, motion pictures—the medium that Akeley believed would enhance museum taxidermy displays—have grown more popular than he ever
imagined, almost completely usurping the popularity enjoyed by museum
taxidermy at the turn of the twentieth century. Hummingbirds under domed
glass bell jars in the Victorian parlor have been replaced by large mammals
of the African savannah on television, in the theatre, on computer screens,
and on mobile devices from the home to the classroom. Given that much of
the world's wildlife is now familiar to the average museum visitor, exhibit
designers are challenged to create ever more entertaining animal displays.
The Behring Family Hall of Mammals exemplifies the modern natural history museum exhibit environment. Its unifying educational narrative, "the
history of mammals," is written in hundreds of taxidermy mounts, thousands of wildlife photographs, and dozens of screens playing videos of the
animals in their natural habitats, in the sound of thunder simulating a rain-
storm on the African savannah, and in the cool air of the Arctic zone.[10]
Along the walls of the exhibit hall hang photographs of various mammalian
species—as though they were portraits of our own family and ancestors—
compelling the museum visitor to gaze into nature's mirror.

ACKNOWLEDGMENTS

I would like to thank the helpful research staff of the many museums, archives, and special collections that I visited in the process of researching and writing this book: special thanks to Pamela Henson at the Smithsonian Institution Archives; Linda Gordon at the Smithsonian Institution National Museum of Natural History, Division of Mammals; Catharine Hawks, private conservator and consultant to the Smithsonian; Ingrid Lennon at the American Museum of Natural History, Department of Library Services; Bernadette Callery at the Carnegie Museum of Natural History Library and Suzanne McLaren, Section of Mammals, Carnegie Museum of Natural History; Judith Chupasko at the Harvard University Museum of Comparative Zoology, Department of Mammalogy; and staff at the Field Museum of Natural History Library; the Ernst Mayr Library, Special Collections, Museum of Comparative Zoology; the Library of Congress Reading Room; and the Department of Rare Books and Special Collections, Rush Rhees Library, University of Rochester. Thanks also to Lynn Nyhart and Robert Kohler for sharing their book manuscripts with me when they were still in their unpublished form.

Parts of this book appeared in earlier versions as two journal-length articles: Chapter 2 appeared as "Breathing New Life into Stuffed Animals: The Society of American Taxidermists, 1880–1885," *Collections: A Journal for Museum and Archives Professionals* 1, no. 2 (November 2004): 155–201. Parts of chapter 3 were reprinted from *Endeavour* 29, no. 3 (September 2005), Mary Anne Andrei, "The Accidental Conservationist: William T. Hornaday, the Smithsonian Bison Expeditions and the US National Zoo," 109–13, © 2005, with permission from Elsevier. Special thanks to Henry Nicholls and Arthur Wadsworth for their encouragement and keen editing.

Thanks to Sally Gregory Kohlstedt, Paul Farber, Jennifer Gunn, Jennifer

Alexander, and Mark Borello for their critical comments that helped shape this book. I am especially grateful to friends and colleagues for reading and commenting on my manuscript: William W. Abbot, Paul Brinkman, Karen Parshall, and Karen Rader. I am especially grateful to Heather Ewing for her close editing of the final manuscript while on a Smithsonian travel adventure, for the long talks, and for inspiration. Many thanks are due to Christie Henry, my first editor at the University of Chicago Press, for believing in this book from the very beginning. For their encouragement along the way and gentle prodding to publish, I am grateful to Jill Friedman, Dimiter Kenarov, Patrick Phillips, Jennifer Wicks, and Elliott D. Woods.

My husband, Ted, and my son, Jack, were great company on the many trips to museums and archives to gather research for this book. All the while, Ted provided unbounded enthusiasm and taught me a great deal about the writing process; in the end, he transformed me into a much better writer. I also owe a great debt of gratitude to my father-in-law, Hugh, who is a distinguished historian of the museum institution, and who provided perpetual encouragement with the oft-spoken phrase "Get to Work!" And to my mother-in-law, Joyce, who not only reiterated those words, but also provided steadfast support over these many years. And I thank my parents, Patricia and Gilberto Andrei, my family, and especially my sister Tammy for her friendship and unwavering belief in me. I couldn't have done it without all of you.

NOTES

INTRODUCTION

1. Frederic A. Lucas, *Fifty Years of Museum Work: Autobiography, Unpublished Papers, and Bibliography* (New York: American Museum of Natural History, 1933), 12.

2. "Africa Transplanted," *Time*, June 1, 1936, 52.

3. "Akeley Memorial Dedicated by 2,000," *New York Times*, March 20, 1936, 3.

4. Russell Owen, "Africa Comes to Life in New York," *New York Times*, May 17, 1936, SM16.

5. The following description of Akeley's first encounter with Henry A. Ward and quoted passages, unless otherwise noted, are taken from C. E. Akeley, *In Brightest Africa*, memorial ed. (Garden City, NY: Garden City Publishing, [1923] 1930), 3–4.

6. Akeley, *In Brightest Africa*, 5–6.

7. Akeley, *In Brightest Africa*, 5–6.

8. Frederic A. Lucas, "Akeley as a Taxidermist," *Natural History* 27 (March–April 1927): 145; Frederic A. Lucas, "Glimpses of Early Museums," *Natural History*, January–February 1921, 77, 75.

9. Lucas, "Akeley as a Taxidermist," 145.

10. Paul Farber, "The Development of Taxidermy and the History of Ornithology," *Isis* 68, no. 244 (1977): 552n.

11. "Divers Means for preserving from Corruption dead Birds, that they may arrive there in good Condition. Some of the same Means may be employed for preserving Quadrupeds, Reptiles, Fishes, and Insects," quoted in Farber, "Development of Taxidermy," 550–53; for discussion of Réaumur's elephant, see Louise E. Robins, *Elephant Slaves and Pampered Parrots: Exotic Animals in Eighteenth-Century Paris* (Baltimore: Johns Hopkins University Press, 2002), 19.

12. Jean-Baptiste Bécoeur, "Fin de la replique de M. Bécoeur à la letter de M. Mauduyt, insérée dans le *Journal de physique* du mois de Novembre 1774," *Journal encyclopédique* 5 (1775): 143, quoted in Farber, "Development of Taxidermy," 557.

13. Farber, "Development of Taxidermy," 550–66; Stephen L. Williams and Catharine A. Hawks, "History of the Preparation Materials Used for Recent Mammal Specimens," in *Mammal Collection Management*, ed. Hugh H. Genoways, Clyde Jones, and Olga L.

Rossolimo (Lubbock: Texas Tech University Press, 1987), 21–49; Charles Willson Peale, *The Autobiography of Charles Willson Peale*, ed. Stanley Hart, vol. 5 of *The Selected Papers of Charles Willson Peale and His Family*, ed. Lillian Miller, Sidney Hart, David C. Ward, Lauren E. Brown, Sara C. Hale, and Leslie K. Reinhardt (New Haven: Yale University Press for the National Portrait Gallery, Smithsonian Institution, 2000), 223.

14. Peale, *Autobiography*, 309.

15. Lucas, "Akeley as a Taxidermist," 145.

16. William Bullock, *A Concise and Easy Method of Preserving Subjects of Natural History* (London, 1817), quoted in S. Peter Dance, *The Art of Natural History* (Woodstock, NY: Overlook Press, 1978), 127.

17. William Bullock, *A Companion to the London Museum and Pantherion* (London, 1813), 136, 109–10.

18. Bullock, *Companion*, iv.

19. Miquel Molina, "More Notes on the Verreaux Brothers," *Pula: Botswana Journal of African Studies* 16 (2002): 30–36.

20. Frederic A. Lucas, "The Story of Museum Groups," *American Museum Journal*, January 1914, 10.

21. Ernst Mayr, *Animal Species and Evolution* (Cambridge, MA: Belknap Press of Harvard University Press, 1963).

22. W. Z. Lidicker Jr., "The Nature of the Subspecific Boundaries in a Desert Rodent and Its Implications for Subspecific Taxonomy," *Systematic Zoology* 11 (1962): 160–71; Hugh H. Genoways, "Philosophy and Ethics of Museum Collection Management," in *Proceedings of the Workshop on Management of Mammal Collection in Tropical Environments, Held at Calcutta from 19th to 25th January, 1984* (Zoological Survey of India, 1988), 9; Robert E. Kohler, "Subspecies Classification and Biological Survey, 1850s–1930s," Max-Planck-Institut für Wissenschaftsgeschichte Preprint Series, no. 240 (2003), 31–32, since revised into chap. 6 of *All Creatures: Naturalists, Collectors, and Biodiversity, 1850–1950* (Princeton, NJ: Princeton University Press, 2006); Williams and Hawks, "History of the Preparation Materials Used," 22. Mary P. Winsor has clearly demonstrated that it was John Edward Gray of the British Museum of Natural History who set the precedent for this change when, in 1864, he proposed to create a scientific and educational division in the collections of the British Museum. See Mary P. Winsor, "Agassiz's Notions of a Museum: The Vision and the Myth," in *Cultures and Institutions of Natural History: Essays in the History and Philosophy of Science*, ed. Michael T. Ghiselin and Alan E. Leviton (San Francisco: California Academy of Sciences, 2000), 249–71.

23. On the history and cultural significance of zoos in America, see Elizabeth Hanson, *Animal Attractions: Nature on Display in American Zoos* (Princeton, NJ: Princeton University Press, 2002), chap. 5 ("Natural Settings").

24. See Mark V. Barrow Jr., "The Specimen Dealer: Entrepreneurial Natural History in America's Gilded Age," *Journal of the History of Biology* 33 (2000): 502; and Sally Gregory Kohlstedt, "Henry A. Ward: The Merchant Naturalist and American Museum Development," *Journal of the Society for the Bibliography of Natural History* 9, no. 4 (1980): 647.

25. William T. Hornaday, "Eighty Fascinating Years" (unpublished autobiography), n.d., Miscellaneous drafts, container 17, William T. Hornaday Papers, Library of Congress, chap. 3, 26.

26. John F. Reiger, *American Sportsmen and the Origins of Conservation*, 3rd ed. (Corvallis: Oregon State University Press, 2001), 46.

27. Reiger, *American Sportsmen*, 6–9.

28. William T. Hornaday, *Our Vanishing Wild Life* (New York: New York Zoological Society, 1912), 54.

29. Hornaday, *Our Vanishing Wild Life*, 54.

30. Hornaday, *Our Vanishing Wild Life*, 56.

31. Hornaday, *Our Vanishing Wild Life*, 117.

32. Hornaday, *Our Vanishing Wild Life*, 101.

33. Hornaday, *Our Vanishing Wild Life*, 384.

34. Hornaday, *Our Vanishing Wild Life*, 105.

35. Hornaday, *Our Vanishing Wild Life*, 278.

36. Hornaday, *Our Vanishing Wild Life*, 396.

37. Hornaday, *Our Vanishing Wild Life*, 392.

38. Douglas Brinkley, "Frontier Prophets," *Audubon*, November–December 2010.

CHAPTER ONE

1. Charles Haskins Townsend, "In Memoriam: Frederic Augustus Lucas," *The Auk* 47 (1930): 147–48.

2. Hornaday, "Eighty Fascinating Years," chap. 3, 25. For a biography of Hornaday, see James A. Dolph, "Bringing Wildlife to Millions: William Temple Hornaday: The Early Years, 1854–1896" (PhD diss., University of Massachusetts, 1975).

3. E. S. Morse, "Notes," *American Naturalist*, April 1873, 250, 252; Hornaday, "Eighty Fascinating Years," chap. 2, 24.

4. Hornaday, "Eighty Fascinating Years," chap. 2, 23, 24; Roswell Howell Ward, *Henry A. Ward: Museum Builder to America* (Rochester, NY: Rochester Historical Society, 1948), 173–74.

5. Hornaday, "Eighty Fascinating Years," chap. 3, 25.

6. Ward, *Henry A. Ward*, 34.

7. Ward, *Henry A. Ward*, 123.

8. Ward, *Henry A. Ward*, 130.

9. Ward, *Henry A. Ward*, 131.

10. Ward, *Henry A. Ward*, 157–58.

11. Lucas, *Fifty Years*, 9; Ward, *Henry A. Ward*, 163.

12. Lucas, *Fifty Years*, 11.

13. Ward, *Henry A. Ward*, 163; Lucas, *Fifty Years*, 10; Townsend, "In Memoriam," 148.

14. Ward, *Henry A. Ward*, 172.

15. Lucas, *Fifty Years*, 11, 12.

16. Townsend, "In Memoriam," 148.

17. The following description of Hornaday's arrival at Ward's is from Hornaday, "Eighty Fascinating Years," chap. 3, 27.

18. Hornaday, "Eighty Fascinating Years," chap. 3, 27.

19. Hornaday, "Eighty Fascinating Years," chap. 2, 19, and chap. 3, 27. Hornaday writes that his first bird specimen was a "Great White Pelican," but the great white

pelican is a species of southeastern Europe, Asia, and Africa; given that Hornaday's specimen was collected on the Iowa State campus in Ames, Iowa, this bird must have been an American white pelican.

20. Hornaday, "Eighty Fascinating Years," chap. 3, 29.

21. Hornaday, "Eighty Fascinating Years," chap. 4, 32.

22. Hornaday, "Eighty Fascinating Years," chap. 4, 36, 39. This specimen created a stir in the American scientific community, as it was the first American crocodile collected for science. Hornaday wrote an article in the *American Naturalist* describing what he believed was a new species, *Crocodilus floridanus*. See William T. Hornaday, "The Crocodile in Florida," *American Naturalist* 9.9 (September 1875): 498–504. Hornaday's Florida crocodile was later determined to be a junior synonym of the species *Crocodylus acutus*, found throughout Central and South America, with Biscayne Bay as its northernmost range—a species first described by celebrated naturalist Georges Cuvier in 1807.

23. Hornaday, "Eighty Fascinating Years," chap. 4, 47–49.

24. Dolph, "Bringing Wildlife to Millions," 79.

25. Stefan Bechtel, *Mr. Hornaday's War: How a Peculiar Victorian Zookeeper Waged a Lonely Crusade for Wildlife That Changed the World* (Boston: Beacon Press, 2012), 100.

26. Dolph, "Bringing Wildlife to Millions," 140.

27. Elliott Coues, "Notice of Mrs. Maxwell's Exhibit of Colorado Mammals," in *On the Plains and Among the Peaks; or, How Mrs. Maxwell Made Her Natural History Collection*, by Mary Emma Dartt Thompson and Martha Maxwell (Philadelphia: Claxton, Remsen, Haffelfinger, 1878), 217. For a biography of Martha Maxwell, see Maxine Benson, *Martha Maxwell: Rocky Mountain Naturalist* (Lincoln: University of Nebraska Press, 1986).

28. "Our Centennial Letters—No. 8," *Forest and Stream*, August 3, 1876, 423.

29. William T. Hornaday, *Two Years in the Jungle* (New York: C. Scribner's Sons, 1886), 4.

30. "In Search of Science," *St. Louis Globe-Democrat*, June 15, 1879, 11; Hornaday, *Two Years in the Jungle*, 2; William T. Hornaday, *Taxidermy and Zoological Collecting* (New York: C. Scribner's Sons, 1902), 9–10 (see this source for a complete list of Hornaday's kit).

31. W. T. Hornaday to H. A. Ward, November 27, 1877, HAWP; Hornaday, *Two Years in the Jungle*, 202.

32. H. A. Ward to Alexander Agassiz, March 31, 1878, AAMDP; W. T. Hornaday to H. A. Ward, December 3, 1877–January 13, 1878, HAWP.

33. Hornaday, *Two Years in the Jungle*, 217. The Museum of Comparative Zoology discarded the specimen in 1973. See the specimen catalogues, MCZ, Department of Mammalogy.

34. Frederic S. Webster, "The Birth of Habitat Groups: Reminiscences Written in His Ninety-Fifth Year," *Annals of the Carnegie Museum* 30 (1945): 101–2.

35. Ruthven Deane, "Extracts from the Field Notes of George B. Sennett," *The Auk*, October 1923, 632.

36. Webster, "Birth of Habitat Groups," 102.

37. Webster, "Birth of Habitat Groups," 104.

38. Webster, "Birth of Habitat Groups," 105.

39. "Taxidermy at Home," *Ward's Natural Science Bulletin*, January 1, 1883, 13.

40. Charles Haskins Townsend, "Old Times with the Birds: Autobiographical with Two Portraits," *The Condor* 29, no. 5 (September–October 1927): 224–32.

41. "In Search of Science," *St. Louis Globe-Democrat*, June 15, 1879, 11.

42. Hornaday, *Two Years in the Jungle*, 371.

43. Hornaday, *Taxidermy and Zoological Collecting*, 230.

44. Hornaday, "Eighty Fascinating Years," chap. 8.

45. Hornaday, Taxidermy and Zoological Collecting, 230.

46. Hornaday, Taxidermy and Zoological Collecting, 230.

47. George Brown Goode, *United States National Museum Annual Report for 1893* (Washington, DC: Government Printing Office, 1895), 42.

48. Hornaday, Taxidermy and Zoological Collecting, 231.

49. Hornaday, Taxidermy and Zoological Collecting, 231.

50. Hornaday, *Taxidermy and Zoological Collecting*, 231; Hornaday, *Two Years in the Jungle*, 371.

51. Hornaday, *Taxidermy and Zoological Collecting*, 233; *Washington Post*, August 17, 1883, 1.

52. Webster, "Birth of Habitat Groups," 106.

53. *The First Annual Report of the Society of American Taxidermists, 1880–81* (Rochester, NY: Daily Democrat and Chronicle Book and Job Print, 1881), 24.

54. *First Annual Report of the Society of American Taxidermists*, 13, 24.

55. *First Annual Report of the Society of American Taxidermists*, 13.

56. In the 1850s and 1860s, a loosely affiliated group of German naturalists had undertaken a similar movement in Germany that they dubbed "practical natural history." At the forefront was taxidermist Philipp Leopold Martin, who was a proponent of "natural" and "lively" biological group displays. For an in-depth discussion of this movement, see Lynn K. Nyhart, *Modern Nature: The Rise of the Biological Perspective in Germany* (Chicago: University of Chicago Press, 2009).

57. *First Annual Report of the Society of American Taxidermists*, 17.

CHAPTER TWO

1. William T. Hornaday, *A Wild Animal Round-Up* (New York: C. Scribner's Sons, 1925), 299.

2. "Our Rochester Letter," *Forest and Stream*, December 23, 1880, 409; "The First Taxidermists' Exhibition," correspondence of the *New York Tribune*, 1880, reprinted in *Ward's Natural Science Bulletin* 1, no. 1 (June 1, 1881): 14.

3. "First Taxidermists' Exhibition," 14.

4. "First Taxidermists' Exhibition," 14; Hornaday, *Taxidermy and Zoological Collecting*, 223.

5. "First Taxidermists' Exhibition," 14.

6. Webster, "Birth of Habitat Groups," 108.

7. *First Annual Report of the Society of American Taxidermists*, 10.

8. Webster, "Birth of Habitat Groups," 107.

9. Webster, "Birth of Habitat Groups," 107.

10. Webster, "Birth of Habitat Groups," 107.

11. John J. Audubon, *The Birds of America: From Drawings Made in the United States and Their Territories,* vol. 6 (Philadelphia: John Bowen, 1843), 143.

12. Frank M. Chapman, "A Flamingo City: Recording a Recent Exploration into a Little-Known Field of Ornithology," *Century* 69, no. 2 (December 1904): 164; "Notes and News," *The Auk* 22 (1905): 109. This article describes an elaborate diorama of nesting American flamingos, designed by Chapman and installed in the AMNH in 1905.

13. In *Habitat Dioramas: Illusions of Wilderness in Museums of Natural History* (Stockholm, Sweden: Almqvist & Wiksell, 1993), 18, Karen Wonders erroneously argues that the "judges . . . rejected Webster's flamingo group with its painted background. . . . By contrast, Hornaday's orang display was given the highest award. . . . Although it featured a more sensationalized scene, it had no painted background." "The Flamingo at Home" did not have a painted background until 1884, when Ward added it to the Zoological Series of the Milwaukee collection.

14. Webster, "Birth of Habitat Groups," 108.

15. Hornaday, "Eighty Fascinating Years," chap. 8.

16. Hornaday, "Eighty Fascinating Years," chap. 8.

17. Hornaday, *Taxidermy and Zoological Collecting,* 232.

18. *First Annual Report of the Society of American Taxidermists,* 19.

19. "The Last Night of a Successful Exhibition," *Rochester Democrat and Chronicle,* December 23, 1880, reprinted in *Ward's Natural Science Bulletin* 1, no. 1 (June 1, 1881): 14.

20. *The Second Annual Report of the Society of American Taxidermists, 1881–82* (Rochester, NY: Daily Democrat and Chronicle Book and Job Print, 1882), 34.

21. *Second Annual Report of the Society of American Taxidermists,* 34.

22. *Second Annual Report of the Society of American Taxidermists,* 35.

23. Coues, "Notice of Mrs. Maxwell's Exhibit," 217.

24. *First Annual Report of the Society of American Taxidermists,* 24.

25. *Second Annual Report of the Society of American Taxidermists,* 35–36.

26. *Second Annual Report of the Society of American Taxidermists,* 36–37.

27. *Second Annual Report of the Society of American Taxidermists,* 37.

28. *Second Annual Report of the Society of American Taxidermists,* 37.

29. "Fin, Fur and Feather," *Boston Journal,* December 15, 1881.

30. *Boston Evening Transcript,* December 15, 1881.

31. Reporters from both the *Boston Journal* and the *Boston Evening Transcript* of December 15, 1881, described it as "an Indian elephant two feet nine inches in height, and not more than six or eight months old when it came to its death," leaving little doubt that they were using the exhibition catalogue to write their reports.

32. The SAT exhibition catalogue often uses the term "suddenly" to describe the action taking place in a group. For the taxidermist, it was important that a group convey a definite sense of animation by capturing an unexpected moment in nature, often between predator and prey.

33. Mary E. W. Jeffrey was a member of the SAT and designed two chicken penwipers, two kitten penwipers (presumably to blot a quill pen), and a rug (woodchuck with head mounted) for the second exhibition. By the publication of the *Second Annual Report,* she is listed as deceased: "New York, October 15, 1882."

34. *Second Annual Report of the Society of American Taxidermists,* 10.

35. Hornaday, *Taxidermy and Zoological Collecting*, 220.

36. Hornaday, *Taxidermy and Zoological Collecting*, 221.

37. Owned by Mary E. W. Jeffrey.

38. *Ward's Natural Science Bulletin* 1, no. 2 (January 1, 1882): 13.

39. "The Society of Taxidermists," *Ward's Natural Science Bulletin* 2, no. 1 (January 1, 1883): 2.

40. *Ward's Natural Science Bulletin* 1, no. 3 (April 1, 1882): 1.

41. Figures are generated from the treasurer's reports in the *First Annual Report of the Society of American Taxidermists*, 20; and the *Second Annual Report of the Society of American Taxidermists*, 25–26. The society's expenditures from the second exhibition exceeded its income by $314. Advertising cost was $191.88, compared with only $34.36 spent the year before.

42. *Second Annual Report of the Society of American Taxidermists*, 22–23.

43. Advertisements, *Boston Herald*, December 15 and 16, 1881.

44. Goode, *United States National Museum Annual Report for 1893*, 44.

45. Goode, *United States National Museum Annual Report for 1893*, 42.

46. Lucas, *Fifty Years*, 14.

47. The book was worth $45.

48. W. T. Hornaday to H. A. Ward (on SAT letterhead), October 4–12, 1882, HAWP.

49. W. T. Hornaday to H. A. Ward, October 15–18, 1882, HAWP.

50. W. T. Hornaday to H. A. Ward, October 27, 1882, HAWP.

51. W. T. Hornaday to H. A. Ward, December 2–10, 1882, HAWP.

52. W. T. Hornaday to H. A. Ward, December 14, 1882, HAWP.

53. "The Society of Taxidermists," 2.

54. W. T. Hornaday to H. A. Ward, December 22, 1882, HAWP.

55. Hornaday, "Eighty Fascinating Years," chap. 15.

56. Andrew Carnegie, *Round the World* (New York: Scribner, 1884), 161.

57. *The Third Annual Report of the Society of American Taxidermists, 1882–83* (Rochester, NY: Daily Democrat and Chronicle Book and Job Print, 1883), 35.

58. "Taxidermy as an Art," *New York Times*, May 1, 1883.

59. J. B. Holder, "Address of Dr. J. B. Holder," in *Third Annual Report of the Society of American Taxidermists*, 43.

60. Holder, "Address of Dr. J. B. Holder," 43.

61. Holder, "Address of Dr. J. B. Holder," 43.

62. Holder, "Address of Dr. J. B. Holder," 43.

63. Holder, "Address of Dr. J. B. Holder," 46.

64. Holder, "Address of Dr. J. B. Holder," 43.

65. Frederic A. Lucas, "The Scope and Needs of Taxidermy," in *Third Annual Report of the Society of American Taxidermists*, 53.

66. Laura Wood Roper, *FLO: A Biography of Frederick Law Olmsted* (1973; repr., Baltimore: Johns Hopkins University Press, 1983), 292. Although Olmsted did not agree with Vaux's classification of landscape architecture as a fine art, he presented Vaux's concept to the public and is often credited with its development.

67. Lucas, "Scope and Needs of Taxidermy," 52.

68. Lucas, "Scope and Needs of Taxidermy," 53.

69. Lucas, "Scope and Needs of Taxidermy," 53.

70. Lucas, *Fifty Years of Museum Work*, 12.

71. Lucas, "Scope and Needs of Taxidermy," 56.

72. "American Taxidermists' Exhibition," *Forest and Stream* 20, no. 14 (May 3, 1883): 267.

73. Hornaday, *Taxidermy and Zoological Collecting*, 112.

74. Hornaday, *Wild Animal Round-Up*, 303.

75. The following description of Hornaday's mounting of the elephant Mungo and quoted passages, unless otherwise noted, are taken from Frederic A. Lucas, "The Mounting of Mungo," *Science* 7, no. 193 (1886): 337–41.

76. "The Society of American Taxidermists," *Scientific American* 48, no. 20 (May 19, 1883): 305.

77. Hornaday, *Wild Animal Round-Up*, 303.

78. "Society of American Taxidermists," 305.

79. R. W. Shufeldt, "Scientific Taxidermy for Museums," in *Report of the National Museum for the Year Ending June 30, 1892* (Washington, DC: Government Printing Office, 1893), 426.

80. "The Taxidermists' Exhibition," *Ward's Natural Science Bulletin* 2, no. 2 (April 1, 1883): 13.

81. "American Taxidermists' Exhibition," 267.

82. T. W. Frain, "The Taxidermists' Exhibition," *Forest and Stream* 20, no. 16 (May 17, 1883): 305.

83. "American Taxidermists' Exhibition," 267.

84. Lucas, "Akeley as a Taxidermist," 144.

85. See "Snap Shots," *Journal of Outdoor Life*, January 26, 1893, 1; and C. Hart Merriam, "The Biological Survey—Origin and Early Days—A Retrospect," *Survey* 16, no. 3 (March 1933): 4.

86. "Snap Shots," 1.

87. Lucas, "Akeley as a Taxidermist," 144.

88. Akeley, *In Brightest Africa*, 7–8.

89. *Harper's Weekly*, May 5, 1883.

90. "American Taxidermists' Exhibition," 267.

91. "Our Group of Ornithorhynchus," *Ward's Natural Science Bulletin* 2, no. 2 (April 1, 1883): 9. Although the society's members were not aware of any antecedents to this group, it is important to note that in 1787, Charles Willson Peale displayed a similar aquatic habitat at his Philadelphia Museum on Lombard Street. For a detailed description of the exhibit, see Peale, *Autobiography*, 221–22, 309. For a comparison of Peale's exhibition techniques with those of museums in Europe, see Toby A. Appel, "Science, Popular Culture and Profit: Peale's Philadelphia Museum," *Journal of the Society for the Bibliography of Natural History* 9 (1980): 619–34.

92. "Our Group of Ornithorhynchus," 9.

93. "Our Group of Ornithorhynchus," 9.

94. "The Taxidermists' Exhibition," 13.

95. "The Taxidermists' Exhibition," 13.

96. "The Taxidermists' Exhibition," 13.

97. Hornaday, *Taxidermy and Zoological Collecting*, 249–50.

98. *Third Annual Report of the Society of American Taxidermists*, 11.

99. Hornaday, *Taxidermy and Zoological Collecting*, 249–50.

100. *Third Annual Report of the Society of American Taxidermists*, 112–14.

101. "Revolution in Taxidermy," *New York Commercial Advertiser*, May 3, 1883, reprinted in *Ward's Natural Science Bulletin* 2, no. 2 (April 1, 1883): 16.

102. "The Society of Taxidermists," 2.

103. *Third Annual Report of the Society of American Taxidermists*, 7.

104. *Annual Report of the Board of Regents of the Smithsonian Institution for the Year 1884* (Washington, DC: Government Printing Office, 1885), 66.

105. W. T. Hornaday, Circular to the members of the Society of American Taxidermists in relation to an exhibit at the New Orleans Exposition, September 17, 1884, SIA, RU 70; E. A. Burke to G. Brown Goode, August 4, 1884, SIA, RU 70; Exposition Reports of the Smithsonian Institution and the United States National Museum, 1867–1940, SIA, RU 70; W. T. Hornaday to F. Webster, September 13, 1884, SIA, RU 210; Taxidermist, United States National Museum, 1883–1889, SIA, RU 210.

106. No records remain to indicate whether or not prizes were awarded, but given that no new mounts were displayed and no SAT members entered anything except through the National Museum, it is reasonable to assume that none were given.

107. Hornaday, Circular to the members.

108. Hornaday, Circular to the members.

109. *National Republican*, May 19, 1884, 4; *Washington Post*, May 19, 1884, 1.

110. *New Orleans Times-Democrat*, January 8, 1885.

111. "The World's Exposition: From a Sportsman's Standpoint," *Forest and Stream* 24, no. 4 (February 19, 1885): 64.

112. Hornaday, "Eighty Fascinating Years," chap. 8; Lucas, *Fifty Years*, 15.

113. "Good and Bad Taxidermal Art," *Scientific American* 55, no. 9 (August 28, 1886): 129.

114. "In a Taxidermist's Studio: How Animal Pets Are Stuffed and Mounted," *Washington Post*, August 10, 1885, 1.

115. Hornaday, "Eighty Fascinating Years," chap. 8.

116. Hornaday, "Eighty Fascinating Years," chap. 8.

117. Hornaday, "Eighty Fascinating Years," chap. 8.

118. Lucas, *Fifty Years*, 15.

CHAPTER THREE

1. Frederic A. Lucas. "Animals Recently Extinct or Threatened with Extermination, as Represented in the Collections of the U.S. National Museum," in *Annual Report of the Board of Regents of the Smithsonian Institution for the Year Ending June 30, 1889* (Washington, DC: Government Printing Office, 1891), 609.

2. *Annual Report of the Board of Regents of the Smithsonian Institution for the Year Ending June 30, 1886* (Washington, DC: Government Printing Office, 1889), 80–81.

3. "List and Classification of the Mounted Mammal Specimens Prepared to Be Exhibited at the New Orleans Exposition, by the U.S. National Museum," May 14, 1884, SIA, RU 70.

4. *Report of the U.S. National Museum under the Direction of the Smithsonian Institution for the Year Ending June 30, 1886* (Washington, DC: Government Printing Office, 1889), 79.

5. William T. Hornaday, "Extermination of the American Bison," in *Annual Report of the Board of Regents of the Smithsonian Institution for the Year Ending June 30, 1887*, 529.

6. Briton Cooper Busch, *The War against the Seals: A History of the North American Seal Fishery* (Kingston: McGill-Queen's University Press, 1985).

7. Charles H. Townsend, "The California Sea-Elephant," *Forest and Stream*, January 13, 1887, 485.

8. This and the following paragraphs, unless otherwise noted, derive their information from Townsend, "California Sea-Elephant," 485; and Charles H. Townsend, "The Elephant Seal Not Extinct," *Century*, June 1912, 205–11.

9. Townsend, "California Sea-Elephant," 485.

10. Townsend, "Elephant Seal Not Extinct," 206.

11. Townsend, "California Sea-Elephant," 485.

12. Townsend, "California Sea-Elephant," 485.

13. Townsend, "California Sea-Elephant," 485.

14. Hornaday, "Extermination of the American Bison," prefatory note, 371.

15. Dan Flores, *American Serengeti: The Last Big Animals of the Great Plains* (Lawrence: University Press of Kansas, 2016), 122–24. Flores uncovered historical evidence that proves General Sheridan was not even present at the trial in which he was supposed to have uttered the oft-repeated phrase: "They are destroying the Indian's commissary. . . . Send them powder and lead . . . but for the sake of a lasting peace, let them kill, skin, and sell until the buffalo is exterminated."

16. Dan O'Brian, *Great Plains Bison* (Lincoln: University of Nebraska Press, 2017), 34.

17. This and the following paragraph derive their information from Hornaday, "Extermination of the American Bison," 530.

18. Hornaday, "Eighty Fascinating Years," chap. 10, 20.

19. Hornaday, "Extermination of the American Bison," 532–34.

20. Hornaday, "Extermination of the American Bison," 396.

21. William T. Hornaday, "The Passing of the Buffalo–I," *Cosmopolitan* 4 (October 1887): 91.

22. Hornaday, "Passing of the Buffalo–I," 92.

23. Hornaday, "Eighty Fascinating Years," chap. 10, 20.

24. *Washington Post*, August 30, 1886, 2.

25. "The National Museum Buffalo," *Forest and Stream* 28, no. 6 (March 1877): 3.

26. Hornaday, "Extermination of the American Bison," 545.

27. Frederic A. Lucas, "The Expedition to Funk Island, with Observations upon the History and Anatomy of the Great Auk," in *Report of the U.S. National Museum for 1888* (Washington, DC: U.S. Government Printing Office, 1890), 493.

28. Lucas, "Animals Recently Extinct, 610.

29. Frederic A. Lucas, "Official Extermination," *Forest and Stream* 28 (March 3, 1887), 104.

30. Lucas, "Official Extermination," 104.

31. Brent S. Stewart and Harriet R. Huber, "*Mirounga angustirostris*," *Mammalian Species* no. 449 (November 15, 1993): 1–10.

32. Lucas, "Expedition to Funk Island," 507.

33. Lucas, "Expedition to Funk Island," 719.

34. Lucas, "Expedition to Funk Island," 512.

35. Lucas, "Expedition to Funk Island," 512.

36. Lucas, "Expedition to Funk Island," 512.

37. Lucas, "Animals Recently Extinct," 613.

38. Hornaday, "Extermination of the American Bison," 522.

39. *Report upon the Condition and Progress of the U.S. National Museum during the Year Ending June 30, 1888* (Washington, DC: Government Printing Office, 1889), 20.

40. Hornaday, "Eighty Fascinating Years," chap. 9, 5; "Buffalo for Washington," *Forest and Stream* 30, no. 13 (April 19, 1888); Hornaday, "Extermination of the American Bison," 463.

41. *Washington Star*, November 12, 1887.

42. Hornaday, *Taxidermy and Zoological Collecting*, 233.

43. Hornaday, *Taxidermy and Zoological Collecting*, 233.

44. Hornaday, *Taxidermy and Zoological Collecting*, 233.

45. Hornaday, *Taxidermy and Zoological Collecting*, 244.

46. "Our Group of Ornithorhynchus," 9.

47. Hornaday, "Extermination of the American Bison," 403.

48. Hornaday, *Taxidermy and Zoological Collecting*, 246.

49. Hornaday, *Taxidermy and Zoological Collecting*, 246.

50. Hornaday, *Taxidermy and Zoological Collecting*, 246.

51. Hornaday, *Taxidermy and Zoological Collecting*, 246.

52. *Washington Star*, March 10, 1888.

53. Hornaday, *Taxidermy and Zoological Collecting*, 244.

54. A letter in the *Chicago News*, later reprinted in "The Breakfast Table," *Boston Daily Advertiser*, April 16, 1887, 4.

55. A letter in the *Chicago News*, later reprinted in "The Breakfast Table," *Boston Daily Advertiser*, April 16, 1887, 4.

56. "Letters to the Editor," *Boston Daily Advertiser*, April 25, 1887, 4.

57. *Boston Daily Advertiser*, April 25, 1887, 4.

58. George Brown Goode, *Annual Report of the Board of Regents of the Smithsonian Institution, and Report of the U.S. National Museum for the Year Ending June 30, 1888* (Washington, DC: Government Printing Office, 1890), 60.

59. R. W. Shufeldt, "Scientific Taxidermy for Museums," in *Report of the National Museum for the Year Ending June 30, 1892* (Washington, DC: Government Printing Office, 1893), 422.

60. Paulus Potter (1625–1654) was a Dutch painter of landscapes and animals. Shufeldt, "Scientific Taxidermy for Museums," 423.

61. "A Story of Destruction," *Forest and Stream*, 31, no. 10 (1888): 181.

62. "A Story of Destruction," 181.

63. Susan Leigh Star, "Craft vs. Commodity, Mess vs. Transcendence: How the Right Tool Became the Wrong One in the Case of Taxidermy and Natural History," in *The*

Right Tools for the Job: At Work in Twentieth-Century Life Sciences, ed. Adele Clarke and Joan H. Fujimura (Princeton, NJ: Princeton University Press, 1992), 281.

64. *Report of the U.S. National Museum under the Direction of the Smithsonian Institution for the Year Ending June 30, 1889* (Washington, DC: Government Printing Office, 1891), 166.

65. Hornaday, "Extermination of the American Bison," 528.

66. For an in-depth discussion of the establishment of the National Zoo, see Vernon M. Kisling, Jr., "The Origin and Development of American Zoological Parks to 1899," and Helen Lefkowitz Horowitz, "The National Zoological Park: 'City of Refuge' or Zoo?," in *New Worlds New Animals: From Menagerie to Zoological Park in the Nineteenth Century*, ed. Robert J. Hoage and William A. Deiss (Baltimore: Johns Hopkins University Press, 1996). For a detailed account of the history and cultural significance of zoos in America, see Hanson, *Animal Attractions*.

67. Hornaday, "Eighty Fascinating Years," chap. 9, 16.

68. Hornaday, "Eighty Fascinating Years," chap. 11; Dolph, *Bringing Wildlife to Millions*, 668.

69. Hornaday, "Eighty Fascinating Years," chap. 11.

70. "New York's Splendid Zoo," *New York Times*, May 3, 1896, 28.

71. Townsend, "In Memoriam"; David Starr Jordan, *The Days of a Man: Being Memories of a Naturalist, Teacher and Minor Prophet of Democracy*, vol. 1 (New York: World Book, 1922), 552.

72. Jordan, *Days of a Man*, 611.

CHAPTER FOUR

1. Frederic S. Webster, "Taxidermy as a Decorative Art," in *Third Annual Report of the Society of American Taxidermists*, 62.

2. For an overview of modern museum development and building design, see Sally Gregory Kohlstedt and Paul Brinkman, "Framing Nature: The Formative Years of Natural History Museum Development in the United States," *Proceedings of the California Academy of Sciences* 55, no. 2, suppl. 1 (September 30, 2004): 7–33. Also see Steven Conn, *Museums and American Intellectual Life, 1876–1926* (Chicago: University of Chicago Press, 1998), chap. 2.

3. W. T. Hornaday to H. A. Ward, May 1882, HAWP.

4. Webster, "Taxidermy as a Decorative Art," 62–63.

5. Webster, "Taxidermy as a Decorative Art," 62–63.

6. Webster, "Taxidermy as a Decorative Art," 62–63.

7. Webster, "Taxidermy as a Decorative Art," 62.

8. *Washington Post*, May 18, 1891, 4; "Jackson's War Horse," *Washington Post*, March 5, 1887, 4; "The President's Deer," *Washington Post*, October 19, 1886, 2.

9. Akeley, *In Brightest Africa*, 15.

10. Akeley, *In Brightest Africa*, 9.

11. Akeley, *In Brightest Africa*, 7.

12. William Morton Wheeler, "Carl Akeley's Early Work and Environment," *Natural History*, 27, no. 2 (March–April 1927), 138.

13. Wheeler, "Carl Akeley's Early Work and Environment," 135.

14. Wheeler, "Carl Akeley's Early Work and Environment," 138.

15. The description of Akeley's method is from Akeley, *In Brightest Africa*, 10–13.

16. Akeley, *In Brightest Africa*, 10–13.

17. William Morton Wheeler, *Seventh Annual Report of the Board of Trustees of the Public Museum of the City of Milwaukee, October 1, 1889* (Milwaukee: Milwaukee Public Museum, 1890), 9.

18. Wheeler, "Carl Akeley's Early Work and Environment," 138.

19. Wheeler, "Carl Akeley's Early Work and Environment," 139.

20. Penelope Bodry-Sanders, *African Obsession: The Life and Legacy of Carl Akeley*, 2nd ed. (Jacksonville, FL: Batax Museum Publishing, 1998), 41.

21. E. Hough, "Elk in Wisconsin," *Forest and Stream*, May 11, 1895, 369.

22. "Before It Is Too Late," *Forest and Stream*, April 11, 1896, 295.

23. *New York Times*, March 4, 1896.

24. See Edward H. McKinley, *The Lure of Africa: American Interests in Tropical Africa, 1919–1939* (Indianapolis: Bobbs-Merrill, 1974), 9–10.

25. Henry Fairfield Osborn, "Report of the President," in *Forty-Fourth Annual Report of the Trustees of the American Museum of Natural History for the Year 1912* (New York: Irving Press, 1913), 16.

26. Henry Fairfield Osborn, "Preservation of the Wild Animals of North America," address before the Boone and Crockett Club (private printing for the Boone and Crockett Club of Washington, 1904); Henry Fairfield Osborn and H. E. Anthony, "The Close of the Age of Mammals," *Journal of Mammalogy* 3, no. 4 (November 1922): 219–37.

27. *Chicago Tribune*, July 19, 1896, 36.

28. *Chicago Tribune*, July 19, 1896, 36.

29. "For the Field Museum," *Chicago Daily Tribune*, November 22, 1896, 1.

30. "For the Field Museum," *Chicago Daily Tribune*, November 22, 1896, 1.

31. "For the Field Museum," *Chicago Daily Tribune*, November 22, 1896, 1.

32. "Unpacking Prof. Elliott's Trophies," *Chicago Daily Tribune*, December 25, 1896, 12.

33. *Annual Report of the Field Museum, for the Year 1897–98* (Chicago, 1898), 193.

34. *Annual Report of the Field Museum, 1897–98*, 193.

35. *Annual Report of the Field Museum, 1897–98*, 193.

36. *Annual Report of the Field Museum, 1897–98*, 193.

37. *Annual Report of the Field Museum, 1897–98*, 193.

38. "Scientific Notes and News," *Science*, December 31, 1897, 991.

39. *Annual Report of the Field Museum, 1897–98*, 287.

40. *Annual Report of the Field Museum, 1897–98*, 365.

41. Bodry-Sanders, *African Obsession*, 68–69, 71.

42. Akeley, *In Brightest Africa*, 11; Bodry-Sanders, *African Obsession*, 69.

43. *Annual Report of the Field Museum, for the Year 1898–99* (Chicago, 1899), 369.

44. Bodry-Sanders, *African Obsession*, 70.

45. The inclusion of two different-aged cubs is not accurate, but mammalogists have only recently made this observation. See Douglas P. DeMaster and Ian Stirling, "*Ursus maritimus*," *Mammalian Species*, no. 145 (May 8, 1981): 1–7.

46. F. S. Webster correspondence with W. T. Hornaday, May 1884–February 1885, Correspondence of the Taxidermist, SIA, RU 210.

47. "In a Taxidermist's Studio," 1.

48. W. J. Holland, *The Carnegie Museum, Annual Report of the Director for the Year Ending March 31, 1898* (Pittsburgh: Murdoch-Kerr Press, 1898), 11.

49. Holland, *Carnegie Museum Annual Report, 1898*, 22.

50. Wapiti, or American elk.

51. Holland, *Carnegie Museum Annual Report, 1898*, 13.

52. Karen Wonders (*Habitat Dioramas*, 165) wrongly interprets Webster's elk group as a result of Hornaday's assertion that Webster's group "suggests with stunning force the steadily accumulating tragedy of 'the Jackson Hole elk herd'" (Hornaday, *Wild Animal Round-Up*, 323–24). However, the range of the California condor, also present in the group, does not extend as far as Yellowstone; also, the elk is depicted as having been killed by a hunter, not by starvation.

53. William L. Finley, "Life History of the California Condor Part II: Historical Data and Range of the Condor," *The Condor* 10, no. 1 (January–February 1908): 5–10.

54. W. J. Holland, *The Carnegie Museum, Annual Report of the Director for the Year Ending March 31, 1900* (Pittsburgh: Murdoch-Kerr Press, 1900), 12.

55. This group is now on display at the National Bird Dog Museum in Grand Junction, Tennessee.

56. Holland, *Carnegie Museum Annual Report, 1900*, 13.

57. Janis C. Sacco and Duane A. Schlitter, "The Return of the Arab Courier: 19th-Century Drama in the North African Desert," *Carnegie Magazine* 62, no. 2 (1994): 31–32, 38–41.

58. Sacco and Schlitter, "Return of the Arab Courier."

59. Holland, *Carnegie Museum Annual Report, 1900*, 12.

60. Holland, *Carnegie Museum Annual Report, 1900*, 13–14.

61. See "Sixteenth Congress of the American Ornithologists' Union," *The Auk*, January 1899, 54.

62. George H. Sherwood, *General Guide to the Exhibition Halls of the American Museum of Natural History*," Guide Leaflet Series of the American Museum of Natural History, no. 35 (New York: American Museum of Natural History, 1911), 64.

63. AOU Minutes, November 17, 1886, AOU Records, SIA, RU 7150, box 5, vol. 1, pp. 251–53. For the full context of this discussion, see Mark V. Barrow Jr., *A Passion for Birds: American Ornithology after Audubon* (Princeton, NJ: Princeton University Press, 1998).

64. *The Carnegie Museum, Annual Report of the Director for the Year Ending March 31, 1901* (Pittsburgh, 1901), 13.

65. "Good Digging in a Wyoming Quarry for the Museum," *Pittsburgh Press*, August 9, 1899.

66. Described from photographs and a pre-1911 postcard of the CMNH bird hall, CMNH Archives.

67. See Barrow, *Passion for Birds*, 133–34; and William Dutcher, "Report of the A. O. U. Committee on the Protection of North American Birds for the Year 1903," *The Auk*, January 1904, 121–24.

68. The group is still on display at the CMNH and is displayed with the original label.

69. Edwin M. Hasbrouck, "The Present Status of the Ivory-Billed Woodpecker (*Campephilus principalis*)," *The Auk*, April 1891, 184.

70. Webster, "Taxidermy as a Decorative Art," 62.

71. Frederick J. V. Skiff to Harlow H. Higinbotham, January 10, 1900, Director's Correspondence—Letterbooks, FMNH Archives.

72. Akeley, *In Brightest Africa*, 14–15.

73. Akeley, *In Brightest Africa*, 15.

74. *Chicago Daily Tribune*, August 14, 1902, 3.

75. F. A. Lucas to C. E. Akeley, April 2, 1902, CEAP-RR.

76. "Museum Notes," *Science*, May 29, 1903, 873–74; F. A. Lucas to C. E. Akeley, December 15, 1902, CEAP-RR.

77. William Alanson Bryan to C. E. Akeley, August 2, 1903, CEAP-RR, 6.2.

78. William Alanson Bryan to C. E. Akeley, August 2, 1903, CEAP-RR, 6.2.

79. William Alanson Bryan to C. E. Akeley, August 2, 1903, CEAP-RR, 6.2.

80. William T. Hornaday to C. E. Akeley, January 11, 1901, CEAP-RR, 3.3.

81. William T. Hornaday to C. E. Akeley, January 11, 1901, CEAP-RR, 3.3.

82. William T. Hornaday, "America Leads the World in the Taxidermist's Art," *New York Herald*, March 21, 1901.

83. Hornaday, "America Leads the World."

84. Harry Denslow to C. E. Akeley, April 14, 1901, CEAP-RR, 3.2.

85. William T. Hornaday to Henry A. Ward, October 15–18, 1882, HAWP.

86. W. T. Hornaday to C. E. Akeley, April 5, 1902, CEAP-RR, 6.1.

87. W. T. Hornaday to C. E. Akeley, February 8, 1901, CEAP-RR, 6.1.

88. Akeley, *In Brightest Africa*, 17–18.

89. Akeley, *In Brightest Africa*, 17–18.

90. Walter L. Beasley, "Modeling Animals in Clay: The Passing of Taxidermy," *Scientific American*, June 25, 1904, 496, 498.

91. W. A. Bryan to C. E. Akeley, July 8, 1904, CEAP-RR, 6.2.

92. W. A. Bryan to C. E. Akeley, July 8, 1904, CEAP-RR, 6.2.

93. W. A. Bryan to C. E. Akeley, July 8, 1904, CEAP-RR, 6.2.

94. Akeley, *In Brightest Africa*, 18.

95. W. A. Bryan to C. E. Akeley, July 8, 1904, CEAP-RR, 6.2.

96. W. A. Bryan to C. E. Akeley, July 8, 1904, CEAP-RR, 6.2.

97. W. A. Bryan to C. E. Akeley, July 8, 1904, CEAP-RR, 6.2.

98. Bodry-Sanders, *African Obsession*, 75, 86, 98.

99. Ward, *Henry A. Ward*, 284.

100. Robert H. Rockwell, *My Way of Becoming a Hunter* (New York: Norton, 1955), 30.

101. *The Carnegie Museum, Annual Report of the Director for the Year Ending March 31, 1907* (Pittsburgh: Murdoch-Kerr Press, 1907), 10.

102. Kenneth C. Parkes, "In Memoriam: Walter Edmond Clyde Todd," *The Auk* 87, no. 4 (October 1987); W. J. Holland, *Carnegie Museum Annual Report, 1900*, 10.

103. Beaming is the process of laying a skin over a beam and scraping the epidermis.

104. Remi Santens to W. J. Holland, November 10, 1906, W. J. Holland Papers, CMNH Archives. Sulphuric acid was used in the pickling process to lower the pH of skins, and salt was used to prevent the acid from swelling the skin.

105. Remi Santens to W. J. Holland, November 10, 1906, W. J. Holland Papers, CMNH Archives.

106. W. E. Clyde Todd to W. J. Holland, January 5, 1907, W. J. Holland Papers, CMNH Archives.

107. *The Carnegie Museum, Annual Report of the Director for the Year Ending March 31, 1908* (Pittsburgh, 1908), 11.

108. Carl E. Akeley, "Address of Carl E. Akeley," in *Theodore Roosevelt: Memorial Addresses Delivered before the Century Association, February 9, 1919* (New York: Century Association, 1919), 62; Thomas Herbert, *Theodore Roosevelt, Typical American: His Life and Work: Patriot, Orator, Historian, Sportsman, Soldier, Statesman and President* (L. H. Walter, 1919), 314; "Woman Killed Elephants," *Washington Post*, February 12, 1907, 13.

109. "Field Museum Plans House for New Jungle Specimens," *Chicago Daily Tribune*, January 30, 1907, 4; "Woman's Bullet Kills Elephant," *Chicago Daily Tribune*, February 10, 1907, 3.

110. "Woman's Bullet Kills Elephant," 3.

111. Bodry-Sanders, *African Obsession*, 100–102.

112. Carl E. Akeley, "An Outline for an Exhibition of the Birds of Illinois," CEAP-RR.

113. Bodry-Sanders, *African Obsession*, 100–102.

114. *Proceedings of the American Association of Museums*, vol. 1, *Records of the Meeting Held at the Museum of the Carnegie Institute, June 4–6, 1907* (Pittsburgh: The Association, 1908).

115. *Proceedings of the American Association of Museums*, vol. 2, *Records of the Third Annual Meeting Held at Chicago, Illinois, May 5–7, 1908* (Charleston, SC: The Association, 1908).

116. *Proceedings of the American Association of Museums*, vol. 2, 1908, 57.

117. *Proceedings of the American Association of Museums*, vol. 2, 1908, 58.

118. *Chicago Daily Tribune*, October 9, 1907, 6.

119. *Chicago Tribune*, July 25, 1909, 5.

120. Hornaday, *Wild Animal Round-Up*, 311.

121. "Scientific Notes and News," *Science*, September 3, 1909, 305.

122. Palle B. Petterson, *Cameras into the Wild: A History of Early Wildlife and Expedition Filmmaking, 1895–1928* (Jefferson, NC: McFarland, 2011), 155.

123. Bodry-Sanders, *African Obsession*, 106.

124. Theodore Roosevelt, *African Game Trails: An Account of the African Wanderings of an American Hunter-Naturalist* (New York: Charles Scribner's Sons, 1910), 399–404; Bodry-Sanders, *African Obsession*, 124–27.

125. Akeley, *In Brightest Africa*, 19.

CHAPTER FIVE

1. Hornaday, *Our Vanishing Wild Life*, ix.

2. "Along Ocean Bottom," *Washington Post*, February 21, 1904, 6.

3. "Along Ocean Bottom," 6.

4. "Deep Sea Monsters," *Washington Post*, March 13, 1904, B6.

NOTES TO PAGES 137–150   225

5. See Frederic A. Lucas, *The Story of Museum Groups*, Guide Leaflet Series, no. 53 (New York: American Museum of Natural History, 1921), 20.

6. Lucas, *Fifty Years*, 26.

7. "Taking the Whale's Cast," *Washington Post*, May 3, 1903, A7.

8. Quoted in Frederick W. True, "The Exhibition of Cetaceans by Papier Maché Casts," *Science* 8, no. 186 (July 22, 1898): 109.

9. *Annual Report of the Board of Regents of the Smithsonian Institution for the Year 1882* (Washington, DC: Government Printing Office, 1884), 125.

10. True, "The Exhibition of Cetaceans by Papier Maché Casts," 109.

11. "Preparing the Big Whale," *Washington Post*, August 16, 1903, E1.

12. "In a Car of Its Own," *Washington Post*, March 13, 1904, E12.

13. Mark Bennitt, ed., *History of the Louisiana Purchase Exposition: Comprising the History of the Louisiana Territory, the Story of the Louisiana Purchase and a Full Account of the Great Exposition, Embracing the Participation of the States and Nations of the World, and Other Events of the St. Louis World's Fair of 1904* (Saint Louis: Universal Exposition Publishing Company, 1905), 321–40.

14. McGrath is quoted in Frederic A. Lucas, "The Passing of the Whale," supplement to the *Zoological Society Bulletin*, July 1908, 446.

15. Lucas, "The Passing of the Whale," 446.

16. Lucas, "The Passing of the Whale," 448.

17. Charles H. Townsend, "The Fate of the Whale," *Forest and Stream*, August 8, 1908, 2.

18. Charles H. Townsend, editor's note on "The Passing of the Whale," supplement to the *Zoological Society Bulletin*, July 1908, 448.

19. Henry Fairfield Osborn, "The New York Zoological Park and Aquarium," *Science* 17, no. 424 (February 13, 1903): 265.

20. Townsend, "Old Times with the Birds," 224–32.

21. "Comforts of Home for Aquarium Fishes," *New York Times Sunday Magazine*, June 5, 1904, 3.

22. "Comforts of Home for Aquarium Fishes," 3.

23. "Overhauling the Aquarium," *New York Times*, December 20, 1903, 24.

24. "Comforts of Home for Aquarium Fishes," 3.

25. "Comforts of Home for Aquarium Fishes," 3.

26. See Charles H. Townsend Correspondence, 1903, NYZS [98-001].

27. "Sewage Killing Off Fish in the Harbor," *New York Times*, July 12, 1908, 5.

28. "Anglers Ready to Act," *New York Times*, March 22, 1908, S4.

29. "Anglers Ready to Act," S4.

30. Charles H. Townsend, "Endurance of the Porpoise in Captivity," *Science* 43, no. 1111 (1916): 534–35.

31. Charles H. Townsend, "West Indian Seals at New York Aquarium," *Forest and Stream*, September 4, 1909, 372.

32. Lucas, *Fifty Years*, 28–29.

33. Frederic A. Lucas, "The Question of Groups," *Museum News*, April 1909, 97–98; Lucas, *Fifty Years*, 27.

34. Frederic A. Lucas, *Museums of the Brooklyn Institute of Arts and Sciences: Report upon the Conditions and Progress of the Museums for the Year Ending December 31, 1905* (New York: Brooklyn Institute, 1906), 12.

35. Lucas, *Museums of the Brooklyn Institute, 1905,* 12.

36. Frederic A. Lucas, *Museum News,* November 1905, 46.

37. *Proceedings of the American Association of Museums,* vol. 1, 73.

38. Lucas found that painted backgrounds could not be painted "for less than fifteen hundred dollars." Lucas, *Museums of the Brooklyn Institute, 1905,* 73.

39. "Scientific Journals and Articles," *Science,* November 10, 1905, 596.

40. *Proceedings of the American Association of Museums,* vol. 1, 70–71.

41. *Proceedings of the American Association of Museums,* vol. 1, 72–73.

42. *Proceedings of the American Association of Museums,* vol. 1, 73.

43. Lucas recounts this anecdote in "Curator F. A. Lucas on His Future Work," *New York Times,* May 14, 1911, 9. Interestingly, Lucas recounted an alternative version of this story four years earlier at the second meeting of the AAM in 1907. See *Proceedings of the American Association of Museums,* vol. 1, 46.

44. Lucas, *Museums of the Brooklyn Institute, 1905,* 12–13.

45. Lucas, *Museum News,* November 1905, 46.

46. "Fur Seal Advisory Board Is Named," *Washington Post,* January 26, 1909, 10.

47. Charles H. Townsend, "Report on Conditions of Fur Seal Herd, 1909," typescript, Records of the New York Aquarium Director's Office, NYZS, RG 7, Control No. 1009.

48. Kurkpatrick Dorsey, *The Dawn of Conservation Diplomacy* (Seattle: University of Washington, 1998), 130; Busch, *War Against the Seals,* 95–98; Karl W. Kenyon and Ford Wilke, "Migration of the Northern Fur Seal, *Callorhinus ursinus,*" *Journal of Mammalogy* 34, no. 1 (February 1953): 86–98.

49. Townsend, "Report on Conditions of Fur Seal Herd, 1909."

50. Townsend, "Report on Conditions of Fur Seal Herd, 1909."

51. William T. Hornaday, *Thirty Years War for Wildlife* (New York: Charles Scribner's Sons for the Permanent Wildlife Protection Fund, 1931), 175.

52. "Wants Seals Protected," *New York Times,* February 27, 1910, 2.

53. Dorsey, *Dawn of Conservation Diplomacy,* 148.

54. Hornaday, *Thirty Years War,* 173–77; "Wants Seals Protected," 2.

55. Frederic A. Lucas, "Breeding Habits of the Pribilof Fur Seal," in *The Fur Seals and Fur-Seal Islands of the North Pacific Ocean* (Washington, DC: Government Printing Office, 1899), 53.

56. Dorsey, *Dawn of Conservation Diplomacy,* 151–52.

57. "Must Die for Race," *Washington Post,* June 30, 1910, 5.

58. W. T. Hornaday to C. Nagel, May 18, 1910, reproduced as "Exhibit C. Letter from the Committee to Secretary Nagel," in Henry W. Elliott, *A Statement Submitted in Re The Fur-Seal Herd of Alaska to the House Committee on the Expenditures in the Department of Commerce* (Washington, DC: Government Printing Office, 1913), 65–66.

59. Hornaday, *Thirty Years War,* 180.

60. "Fur Seal Policy Attacked," *Los Angeles Times,* February 5, 1911, 16; Hornaday, *Thirty Years War,* 180.

61. Townsend, "Elephant Seal Not Extinct," 206.

62. Laurence M. Huey, "Past and Present Status of the Northern Elephant Seal with a Note on the Guadalupe Fur Seal," *Journal of Mammalogy*, 11, no. 2 (May 1930): 188–94; Charles Haskins Townsend, "Voyage of the 'Albatross' to the Gulf of California in 1911," *Bulletin of the American Museum of Natural History* 35 (1916): 405.

63. Townsend, "Voyage of the 'Albatross,'" 405, 407; Townsend, "Elephant Seal Not Extinct," 207.

64. Townsend, "Elephant Seal Not Extinct," 207.

65. Townsend, "Voyage of the 'Albatross,'" 405, 407.

66. "The Newly Discovered Elephant Seals," *Forest and Stream*, April 29, 1911, 652.

67. Townsend, "Voyage of the 'Albatross,'" 405, 407.

68. Townsend, "Voyage of the 'Albatross,'" 405, 407; Townsend, "Elephant Seal Not Extinct," 207.

69. Townsend, "Voyage of the 'Albatross,'" 405, 407.

70. "The Newly Discovered Elephant Seals," 652.

71. Townsend, "Voyage of the 'Albatross,'" 407.

72. Townsend, "Elephant Seal Not Extinct," 207.

73. "Fabled Sea Elephant and Other Rare Creatures Found by Scientists," *New York Times Sunday Magazine*, August 13, 1911, 8.

74. Townsend, "Voyage of the 'Albatross,'" 405, 407.

75. "Sea Elephants on Exhibition," *Forest and Stream*, March 18, 1911, 412.

76. "The Newly Discovered Elephant Seals," 652.

77. Hornaday, *Our Vanishing Wild Life*, 40.

78. Townsend, "Elephant Seal Not Extinct," 207.

79. Hornaday, *Our Vanishing Wild Life*, 40.

80. Townsend, "Elephant Seal Not Extinct," 211.

81. Huey, "Past and Present Status of the Northern Elephant Seal," 188–94.

82. *Fifty-Second Annual Report of the Trustees of the American Museum of Natural History for the Year 1920* (New York: The Museum, 1921), 52.

83. Supplement to the *American Journal of International Law*, vol. 6 (New York: Baker, Voorhis and Company for the Society of International Law, 1912), 162–66.

84. Dorsey, *Dawn of Conservation Diplomacy*, 159.

85. Henry Fairfield Osborn, *Forty-Third Annual Report of the Trustees of the American Museum of Natural History for the Year 1911* (New York: The Museum, 1912), 27.

86. "Dr. Lucas to Head Museum," *New York Times*, May 9, 1911, 20.

87. "Here to Negotiate Treaty: Foreign and American Delegates Plan Protection for Pacific Mammals," *New York Times*, May 11, 1911, 14.

88. "Seal Treaty Signed," *New York Times*, July 8, 1911, 4.

89. Charles H. Townsend, "The Pribilof Fur Seal Herd and the Prospects for Its Increase," *Science* 34, no. 878 (October 27, 1911): 569.

90. Dorsey, *Dawn of Conservation Diplomacy*, 162; Barton W. Evermann. "The Northern Fur-Seal Problem as a Type of Many Problems of Marine Zoology," *Scientific Monthly* 9, no. 3 (September 1919): 271.

91. Frederic A. Lucas, "The Fur Seal Herd," *New York Times*, February 23, 1912, 10.

92. "The Fur-Seal Slaughterers," *New York Times*, February 23, 1912, 10.

93. Townsend. "The Pribilof Fur Seal Herd," 568–70.

94. Marshall McLean, "Discussion and Correspondence: The Pribilof Fur Seal Herd," *Science* 35, no. 892 (February 2, 1912): 183–84.

95. Charles H. Townsend, "Discussion and Correspondence: The Pribilof Fur Seal Herd," *Science* 35, no. 896 (March 1, 1912): 334–36.

96. George A. Clark, "Discussion and Correspondence: The Pribilof Fur Seal Herd," *Science* 35, no. 896 (March 1, 1912): 336–38.

97. Frederic A. Lucas, "The Fur Seal," *American Museum Journal* 12 (1912): 132–33.

98. Lucas, "The Fur Seal," 132–33.

99. Henry Fairfield Osborn, "Preservation of the World's Animal Life," *American Museum Journal* 12 (1912): 124.

100. Wilfred H. Osgood, Edward A. Preble, and George H. Parker, *The Fur Seals and Other Life of the Pribilof Islands, Alaska, in 1914*, Bulletin of the Bureau of Fisheries, vol. 34 (Washington, DC: Government Printing Office, 1914).

101. Henry Fairfield Osborn, foreword to *Our Vanishing Wild Life*, by William T. Hornaday (New York: New York Zoological Society, 1912).

102. Hornaday, *Our Vanishing Wild Life*, 387.

103. Hornaday, *Our Vanishing Wild Life*, 391.

104. In 1921, Joseph Grinnell urged curators of scientific museums to embrace as an ethical principle the careful and accurate preservation of the specimens in their care: "Many species of vertebrate animals are disappearing; some are gone already. All that the investigator of the future will have, to indicate the nature of such then extinct species, will be the remains of these species preserved more or less faithfully, along with the data accompanying them, in the museums of the country." Joseph Grinnell, "The Museum Conscience," *Museum Work* 4 (1921): 62–63.

105. Alja R. Crook, "The Museum and the Conservation Movement," in *Proceedings of the American Association of Museums*, vol. 9, *Records of the Tenth Annual Meeting Held in San Francisco, July 6–9, 1915* (Charleston, SC: The Association, 1915), 93–95.

106. Crook, "The Museum and the Conservation Movement."

107. Crook, "The Museum and the Conservation Movement."

CHAPTER SIX

1. Akeley, *In Brightest Africa*, 251.

2. Akeley, *In Brightest Africa*, 252.

3. "Akeley Back; Has Elephant Jungle," *Chicago Daily Tribune*, November 10, 1911, 1.

4. William T. Hornaday, "Protection of Game for the Gunmakers," *New York Times*, December 20, 1911, 12.

5. Hornaday, *Our Vanishing Wild Life*.

6. Hornaday, *Our Vanishing Wild Life*, 14.

7. Hornaday, *Our Vanishing Wild Life*, 381.

8. "Zoological Society Annual Meeting," *Forest and Stream*, January 20, 1912, 83.

9. Akeley, *In Brightest Africa*, 253.

10. "Akeley Back; Has Elephant Jungle," 1.

11. Akeley, *In Brightest Africa*, 253.

12. Akeley, *In Brightest Africa*, 254.

13. Akeley, *In Brightest Africa*, 6.

14. "Rare Okapi Group on Exhibition Soon," *New York Times*, March 13, 1911, 3.

15. *The Carnegie Museum, Annual Report of the Director for the Year Ending March 31, 1911* (Pittsburgh, 1911), 19.

16. Hornaday, *Wild Animal Round-Up*, 310.

17. The following description of Akeley's plan for the African hall, unless otherwise noted, is taken from a long passage, quoting Akeley's own words, in Mary Cynthia Dickerson, "The New African Hall Planned by Carl E. Akeley: Principles of Construction Which Strike a Revolution in Methods of Exhibition and Presage the Future Greatness of the Educational Museum," *American Museum Journal* 14, no. 5 (May 1914): 180–87.

18. CEAP-AMNH, archive microfilm #76, boxes 1–3.

19. Carl E. Akeley, quoted in Dickerson, "New African Hall," 180–82.

20. Frederic A. Lucas, "General Administration," in *Forty-Fourth Annual Report of the American Museum of Natural History for the Year 1912* (New York: The Museum, 1913), 38.

21. Lucas, "General Administration," 30–31.

22. James L. Clark, *Good Hunting: Fifty Years of Collecting and Preparing Habitat Groups for the American Museum* (Norman, OK: University of Oklahoma Press, 1966), 49.

23. Remi Santens to W. J. Holland, November 10, 1906, WJHP.

24. *Proceedings of the American Association of Museums*, vol. 9, *Records of the Tenth Annual Meeting Held in San Francisco, July 6–9, 1915* (Charleston, SC: The Association, 1915), 105.

25. Hornaday, *Wild Animal Round-Up*, 310.

26. Henry Fairfield Osborn, *Forty-Fifth Annual Report of the American Museum of Natural History for the Year 1913* (New York: The Museum, 1914), 27.

27. Osborn, *Forty-Fifth Annual Report of the AMNH, 1913*, 27.

28. Dickerson, "New African Hall," 180. For another account of the history of African Hall, see Geoffrey Theodore Hellman, *Bankers, Bones and Beetles: The First Century of the American Museum of Natural History* (Garden City, NY: Natural History Press for the American Museum of Natural History, 1969). A popular account of Akeley's African Hall can be read in Joseph Wallace, *A Gathering of Wonders: Behind the Scenes at the American Museum of Natural History* (New York: St. Martin's Press, 2000).

29. Dickerson, "New African Hall," 186.

30. Dickerson, "New African Hall," 186.

31. A. K. Kes Hillman-Smith and Colin P. Groves, "*Diceros bicornis*," *Mammalian Species*, no. 455 (June 2, 1994): 1–8; Colin P. Groves, "*Ceratotherium simum*," *Mammalian Species*, no. 8 (June 16, 1972): 1–6. White rhinos are more gregarious than black rhinos, but Akeley may not have observed this difference. After Akeley's death, James L. Clark, Akeley's assistant, directed the completion of African Hall, at which time the rhinoceroses were not made part of the central floor design, but became one of the habitat groups, depicting the white rhinoceroses as a family group. Another change was in the

addition of four mounts to the elephant group: a young bull was placed in front of the female, another on the left, and a third behind her. The fourth specimen was placed on the old bull's right side. The changes unfortunately altered Akeley's intended narrative.

32. Henry Fairfield Osborn, *Forty-Eighth Annual Report of the American Museum of Natural History for the Year 1916* (New York: The Museum, 1917), 27.

33. Frederic A. Lucas, *Forty-Ninth Annual Report of the American Museum of Natural History for the Year 1917* (New York: The Museum, 1918), 46.

34. Lucas, *Forty-Ninth Annual Report of the AMNH, 1917*, 46–47.

35. Frederic A. Lucas, *Fifty-Third Annual Report of the American Museum of Natural History for the Year 1921* (New York: The Museum, 1922), 31.

36. "Akeley Off to Africa," *New York Times*, July 31, 1921, 67.

37. Lucas, *Fifty-Third Annual Report of the AMNH, 1921*, 44.

38. Frances D. McMullen, "Vivid Hunting Tales Are Told in Bronze," *New York Times*, March 28, 1926, SM5.

39. "Akeley Off to Africa," 67.

40. "Akeley on Gorilla Hunt," *New York Times*, July 18, 1921, 8.

41. "Sailing for Europe to Escape the Heat," *New York Times*, July 30, 1921, 7.

42. Mary L. Jobe Akeley, *Carl Akeley's Africa: The Account of the Akeley-Eastman-Pomeroy African Hall Expedition of the American Museum of Natural History* (New York: Dodd, Mead, 1929), 222.

43. Akeley, *In Brightest Africa*, 264.

44. Akeley, *In Brightest Africa*, 265.

45. Akeley, *In Brightest Africa*, 265.

46. Carl E. Akeley, "Gorillas—Real and Mythical," *Natural History* 23, no. 5 (1923): 437.

47. Mary L. Jobe Akeley, *Carl Akeley's Africa*, 228.

48. Mary L. Jobe Akeley, *Carl Akeley's Africa*, 221–22.

49. William Jennings Bryan, "God and Evolution," *New York Times*, February 26, 1922, sec. 7, 1. For Osborn's response, see Henry Fairfield Osborn, "Evolution and Religion," *New York Times*, March 5, 1922, sec. 7, 2 and 14.

50. Bryan, "God and Evolution."

51. "Akeley Bronze Ape to Stand in Church," *New York Times*, April 9, 1924, 14.

52. "Akeley Bronze Ape to Stand in Church," 14.

53. "Akeley Bronze Ape to Stand in Church," 14.

54. "J. Sultan, Gorilla, Hotel Guest Here," *New York Times*, April 7, 1924, 13; "The 'Almost Human' Gorilla Who Drank Tea and Went to School," *The Guardian*, January 26, 2017.

55. "John Daniel Sniffs at Evolutionists," *New York Times*, April 14, 1924, 20.

56. "Lion a Gentleman, Says Carl Akeley," *New York Times*, April 28, 1924, 10.

57. "The Vanishing Gorilla," *New York Times*, June 26, 1924, 22.

58. Carl E. Akeley, "Have a Heart," *The Mentor*, January 1926, 50.

59. Henry Fairfield Osborn, *Fifty-Seventh Annual Report of the American Museum of Natural History for the Year 1925* (New York: The Museum, 1926), 19.

60. George H. Sherwood (acting director), *Fifty-Eighth Annual Report of the American Museum of Natural History for the Year 1926* (New York: The Museum, 1927), 34.

61. McMullen, "Vivid Hunting Tales Are Told in Bronze," SM5.

62. Sherwood, *Fifty-Eighth Annual Report of the AMNH, 1926*, 34–35.

63. Paul Du Chaillu, *Wild Life under the Equator* (New York: Harper & Brothers, 1869).

64. Akeley, *In Brightest Africa*, 237–38.

65. Akeley, *In Brightest Africa*, 239.

66. Akeley, *In Brightest Africa*, 155.

67. Gregg Mitman, *Reel Nature: America's Romance with Wildlife on Film* (Cambridge, MA: Harvard University Press, 1999), 26–31; Geoff King, Claire Malloy, and Yannis Tzioumakis, *American Independent Cinema: Indie, Indiewood and Beyond* (London: Routledge, 2013), 167–68; Bodry-Sanders, *African Obsession*, 211.

68. Akeley, *In Brightest Africa*, 265.

69. Akeley, *In Brightest Africa*, 263.

70. Akeley, *In Brightest Africa*, 263.

71. Rockwell, *My Way of Becoming a Hunter*, 200.

72. CEAP-AMNH, archive microfilm #76, boxes 1–3.

73. Myles Turner, *My Serengeti Years: The Memoirs of an African Game Warden* (New York: W. W. Norton, 1988), 34a.

74. Rockwell, *My Way of Becoming a Hunter*, 211.

75. Rockwell, *My Way of Becoming a Hunter*, 211.

76. For a complete description of the Akeley method for designing wax vegetation, see Laurence Vail Coleman, *Plants of Wax: How They Are Made in the American Museum of Natural History*, American Museum of Natural History Guide Leaflet Series, no. 54 (February 1928): 1–16.

77. CEAP-AMNH, archive microfilm #76, boxes 1–3.

78. Rockwell, *My Way of Becoming a Hunter*, 213, 227.

79. Carl E. Akeley to H. F. Osborn, October 7, 1925, CEAP-AMNH, archive microfilm #76, boxes 1–3.

80. Carl E. Akeley to H. F. Osborn, October 7, 1925, CEAP-AMNH, archive microfilm #76, boxes 1–3.

81. Rockwell, *My Way of Becoming a Hunter*, 215.

82. "Akeley and African Hall," *New York Times*, December 2, 1926, 26.

83. Rockwell, *My Way of Becoming a Hunter*, 221.

84. Bodry-Sanders, *African Obsession*, 243–44.

85. "Body to Remain in Africa," *New York Times*, December 1, 1926, 12.

86. Bodry-Sanders, *African Obsession*, 249–52.

87. Rockwell, *My Way of Becoming a Hunter*, 263.

88. W. R. Leigh, "Painting the Backgrounds for the African Hall Groups," *Natural History* 30 (1930): 575.

89. W. R. Leigh, *Frontiers of Enchantment: An Artist's Adventures in Africa* (New York: Simon & Schuster, 1938), 32.

90. Rockwell, *My Way of Becoming a Hunter*, 264.

91. Wonders, *Habitat Dioramas*, 176.

92. Wonders, *Habitat Dioramas*, 177.

93. For visitation figures, see the AMNH website at www.amnh.org.

EPILOGUE

1. In 1957, when William L. Brown, then chief taxidermist of the U.S. National Museum, dismantled Hornaday's bison group, he found in the base of the exhibit a metal box containing a copy of Hornaday's article "The Passing of the Buffalo" (*Cosmopolitan*, October 1887), on which Hornaday had written this note. That same year, all five specimens of the group were sent to the University of Montana. They have since been restored and are today on exhibit at the Fort Benton Museum Complex, Montana.

2. Tim Flannery and Peter Schouten, *A Gap in Nature: Discovering the World's Extinct Animals* (New York: Atlantic Monthly Press, 2001).

3. Flannery and Schouten, *A Gap in Nature.*

4. For a complete description of the new hall, see Kathryn M. Duda, "A New Look at North American Wildlife," *Carnegie Magazine* 62, no. 1 (January–February 1995): 28.

5. Duda, "A New Look at North American Wildlife."

6. Catharine Hawks, "Condition Assessment AMNH Akeley Hall of African Mammals, Bird and Mammal Specimens," Conservator's Report (2003). A copy was obtained from the author.

7. Judith Levinson and Sari Uricheck, "Documenting the Documents: The Conservation Survey of the Akeley Hall of African Mammals," *Objects Specialty Group Postprints* 12 (2005): 39–61; Glenn Collins, "Long Live the Elephants, Long Dead," *New York Times*, June 4, 2004.

8. Information regarding the U.S. National Museum's Behring Hall of Mammals is taken from "Kenneth E. Behring Hall of Mammals," in *National Museum of Natural History: Annual Report for 2003*, available at http://www.mnh.si.edu.

9. For an instructive text on museum exhibition and the role of interpretation, see David Dean, *Museum Exhibition: Theory and Practice* (New York: Routledge, 1994).

10. "Kenneth E. Behring Hall of Mammals." For a pictorial account of the Behring Hall taxidermy project, see "Head to Toe: Mammal Makeovers by Smithsonian Taxidermists," *Science in the News*, available at http://www.mnh.si.edu/museum/news /taxidermy.

BIBLIOGRAPHY

MANUSCRIPT COLLECTIONS

AAMDP   Alexander Agassiz, Museum Directors' Papers
Museum of Comparative Zoology—Archives
Ernst Mayr Library, Special Collections

CEAP-AMNH   Carl Ethan Akeley Papers and Mary L. Jobe Akeley Papers
Research Library Special Collections
American Museum of Natural History

CEAP-RR   Carl Ethan Akeley Papers
Rush Rhees Library, Manuscript and Special Collections
University of Rochester

CMNH   Carnegie Museum of Natural History Library

FMNH   Research and Collections Library
Field Museum of Natural History

FSW   Frederic Smith Webster, Biographical Files
Carnegie Museum of Natural History Library

HAWP   The Henry Augustus Ward Papers
Rush Rhees Library, Manuscript and Special Collections
University of Rochester

NYZS   New York Zoological Society Library
Correspondence of Charles H. Townsend, New York Aquarium

RHS   Remi Henri Santens, Biographical Files
Carnegie Museum of Natural History Library

SIA   Smithsonian Institution Archives

WJHP   William Jacob Holland Papers
Carnegie Museum of Natural History Library

WTHP   William T. Hornaday Papers
Library of Congress

PUBLISHED RESOURCES

"Africa Transplanted." *Time*, June 1, 1936, 52.

Akeley, Carl E. "Address of Carl E. Akeley." In *Theodore Roosevelt: Memorial Addresses Delivered Before the Century Association, February 9, 1919.* New York: Century Association, 1919.

———. "Gorillas—Real and Mythical." *Natural History* 23, no. 5 (1923): 437.

———. "Have a Heart." *The Mentor*, January 1926, 50.

———. *In Brightest Africa.* Memorial edition. Garden City, NY: Garden City Publishing, [1923] 1930.

Akeley, Mary L. Jobe. *Carl Akeley's Africa: The Account of the Akeley-Eastman-Pomeroy African Hall Expedition of the American Museum of Natural History.* New York: Dodd, Mead, 1929.

"American Taxidermists' Exhibition." *Forest and Stream* 20, no. 14 (May 3, 1883): 267.

Andrei, Mary Anne. "The Accidental Conservationist: William T. Hornaday, the Smithsonian Bison Expeditions and the U.S. National Zoo." *Endeavour* 29, no. 3 (2005): 109–13.

———. "'Breathing New Life into Stuffed Animals': The Society of American Taxidermists, 1880–1885." *Collections* 1, no. 2 (2004): 155–201.

Annual Report of the Board of Regents of the Smithsonian Institution for the Year 1882. Washington, DC: Government Printing Office, 1884.

Annual Report of the Board of Regents of the Smithsonian Institution for the Year 1884. Washington, DC: Government Printing Office, 1885.

Annual Report of the Board of Regents of the Smithsonian Institution for the Year Ending June 30, 1886. Washington, DC: Government Printing Office, 1889.

Annual Report of the Field Museum, for the Year 1897–98. Chicago, 1898.

Annual Report of the Field Museum, for the Year 1898–99. Chicago, 1899.

Appel, Toby A. "Science, Popular Culture and Profit: Peale's Philadelphia Museum." *Journal of the Society for the Bibliography of Natural History* 9 (1980): 619–34.

Audubon, John James. *The Birds of America: From Drawings Made in the United States and Their Territories.* 7 vols. Philadelphia: John Bowen, 1840–44.

Barrow, Mark V., Jr. *A Passion for Birds: American Ornithology after Audubon.* Princeton, NJ: Princeton University Press, 1998.

———. "The Specimen Dealer: Entrepreneurial Natural History in America's Gilded Age." *Journal of the History of Biology* 33 (2000): 493–534.

Beasley, Walter L. "Modeling Animals in Clay: The Passing of Taxidermy." *Scientific American*, June 25, 1904.

Bechtel, Stefan. *Mr. Hornaday's War: How a Peculiar Victorian Zookeeper Waged a Lonely Crusade for Wildlife That Changed the World.* Boston: Beacon Press, 2012.

"Before It Is Too Late." *Forest and Stream*, April 11, 1896, 295.

Bennitt, Mark, ed. *History of the Louisiana Purchase Exposition: Comprising the History of the Louisiana Territory, the Story of the Louisiana Purchase and a Full Account of the Great Exposition, Embracing the Participation of the States and Nations of the World, and Other Events of the St. Louis World's Fair of 1904*. St. Louis: Universal Exposition Publishing Company, 1905.

Benson, Maxine. *Martha Maxwell: Rocky Mountain Naturalist*. Lincoln: University of Nebraska Press, 1986.

Bodry-Sanders, Penelope. *African Obsession: The Life and Legacy of Carl Akeley*. 2nd ed. Jacksonville, FL: Batax Museum Publishing, 1998.

Brinkley, Douglas. "Frontier Prophets." *Audubon*, November–December 2010.

"Buffalo for Washington." *Forest and Stream* 30, no. 13 (April 19, 1888).

Bullock, William. *A Companion to the London Museum and Pantherion*. London, 1813.

Busch, Briton Cooper. *The War against the Seals: A History of the North American Seal Fishery*. Kingston: McGill-Queen's University Press, 1985.

Carnegie, Andrew. *Round the World*. New York: Scribner, 1884.

The Carnegie Museum, Annual Report of the Director for the Year Ending March 31, 1901. Pittsburgh, 1901.

The Carnegie Museum, Annual Report of the Director for the Year Ending March 31, 1907. Pittsburgh: Murdoch-Kerr Press, 1907.

The Carnegie Museum, Annual Report of the Director for the Year Ending March 31, 1908. Pittsburgh, 1908.

The Carnegie Museum, Annual Report of the Director for the Year Ending March 31, 1911. Pittsburgh, 1911.

Chapman, Frank M. "A Flamingo City: Recording a Recent Exploration into a Little-Known Field of Ornithology." *Century* 69, no. 2 (December 1904): 163–80.

Clark, George A. "Discussion and Correspondence: The Pribilof Fur Seal Herd." *Science* 35, no. 896 (March 1, 1912): 336–38.

Clark, James L. *Good Hunting: Fifty Years of Collecting and Preparing Habitat Groups for the American Museum*. Norman: University of Oklahoma Press, 1966.

Coleman, Laurence Vail. *Plants of Wax: How They Are Made in the American Museum of Natural History*. American Museum of Natural History Guide Leaflet Series, no. 54 (February 1928): 1–16.

"Comforts of Home for Aquarium Fishes." *New York Times Sunday Magazine*, June 5, 1904, 3.

Conn, Steven. *Museums and American Intellectual Life, 1876–1926*. Chicago: University of Chicago Press, 1998.

Coues, Elliott. "Notice of Mrs. Maxwell's Exhibit of Colorado Mammals." In *On the Plains and Among the Peaks; or, How Mrs. Maxwell Made Her Natural History Collection*, by Mary Emma Dartt Thompson and Martha Maxwell. Philadelphia: Claxton, Remsen, Haffelfinger, 1878.

Crook, Alja R. "The Museum and the Conservation Movement." In *Proceedings of the American Association of Museums*, vol. 9, *Records of the Tenth Annual Meeting Held in San Francisco, July 6–9, 1915*, 93–95. Charleston, SC: The Association, 1915.

Dance, S. Peter. *The Art of Natural History*. Woodstock, NY: Overlook Press, 1978.

Dean, David. *Museum Exhibition: Theory and Practice.* New York: Routledge, 1994.

Deane, Ruthven. "Extracts from the Field Notes of George B. Sennett." *The Auk*, October 1923, 632.

DeMaster, Douglas P., and Ian Stirling. "*Ursus maritimus.*" *Mammalian Species*, no. 145 (May 8, 1981): 1–7.

Dickerson, Mary Cynthia. "The New African Hall Planned by Carl E. Akeley: Principles of Construction Which Strike a Revolution in Methods of Exhibition and Presage the Future Greatness of the Educational Museum." *American Museum Journal* 14, no. 5 (May 1914).

Dolph, James A. "Bringing Wildlife to Millions: William Temple Hornaday; The Early Years, 1854–1896." PhD diss., University of Massachusetts, 1975.

Dorsey, Kurkpatrick. *The Dawn of Conservation Diplomacy.* Seattle: University of Washington, 1998.

Du Chaillu, Paul. *Wild Life under the Equator.* New York: Harper & Brothers, 1869.

Duda, Kathryn M. "A New Look at North American Wildlife." *Carnegie Magazine* 62, no. 1 (1995): 28.

Dutcher, William. "Report of the A. O. U. Committee on the Protection of North American Birds for the Year 1903." *The Auk*, January 1904, 121–24.

Elliott, Henry W. "A Statement Submitted in Re The Fur-Seal Herd of Alaska to the House Committee on the Expenditures in the Department of Commerce." Washington, DC: Government Printing Office, 1913.

Evermann, Barton W. "The Northern Fur-Seal Problem as a Type of Many Problems of Marine Zoology." *Scientific Monthly* 9, no. 3 (September 1919): 263–82.

"Fabled Sea Elephant and Other Rare Creatures Found by Scientists." *New York Times Sunday Magazine*, August 13, 1911, 8.

Farber, Paul. "The Development of Taxidermy and the History of Ornithology." *Isis* 68, no. 244 (1977).

Finley, William L. "Life History of the California Condor Part II: Historical Data and Range of the Condor." *The Condor: A Magazine of Western Ornithology* 10, no. 1 (January–February 1908): 5–10.

The First Annual Report of the Society of American Taxidermists, 1880–81. Rochester, NY: Daily Democrat and Chronicle Book and Job Print, 1881.

"The First Taxidermists' Exhibition." Correspondence of the *New York Tribune*, 1880. Reprinted in *Ward's Natural Science Bulletin* 1, no. 1 (June 1, 1881): 14.

Flannery, Tim, and Peter Schouten. *A Gap in Nature: Discovering the World's Extinct Animals.* New York: Atlantic Monthly Press, 2001.

Flores, Dan. *American Serengeti: The Last Big Animals of the Great Plains.* Lawrence: University Press of Kansas, 2016.

Frain, T. W. "The Taxidermists' Exhibition." *Forest and Stream* 20, no. 16 (May 17, 1883): 305.

Genoways, Hugh H. "Philosophy and Ethics of Museum Collection Management." In *Proceedings of the Workshop on Management of Mammal Collection in Tropical Environments, Held at Calcutta from 19th to 25th January, 1984.* Zoological Survey of India, 1988.

"Good and Bad Taxidermal Art." *Scientific American* 55, no. 9 (August 28, 1886): 129.

Goode, George Brown. *Annual Report of the Board of Regents of the Smithsonian Institution, and Report of the U.S. National Museum for the Year Ending June 30, 1888.* Washington, DC: Government Printing Office, 1890.

———. *United States National Museum Annual Report for 1893.* Washington DC: Government Printing Office, 1895.

Grinnell, Joseph. "The Museum Conscience." *Museum Work* 4 (1921): 62–63.

Groves, Colin P. "*Ceratotherium simum.*" *Mammalian Species*, no. 8 (June 16, 1972): 1–6.

Hanson, Elizabeth. *Animal Attractions: Nature on Display in American Zoos.* Princeton, NJ: Princeton University Press, 2002.

Hasbrouck, Edwin M. "The Present Status of the Ivory-Billed Woodpecker (*Campephilus principalis*)." *The Auk*, April 1891, 184.

Hawks, Catharine M. "Condition Assessment AMNH Akeley Hall of African Mammals, Bird and Mammal Specimens." Conservator's Report (2003).

"Head to Toe: Mammal Makeovers by Smithsonian Taxidermists." *Science in the News.* Available at http://www.mnh.si.edu/museum/news/taxidermy.

Hellman, Geoffrey Theodore. *Bankers, Bones and Beetles: The First Century of the American Museum of Natural History.* Garden City, NY: Natural History Press for the American Museum of Natural History, 1969.

Herbert, Thomas. *Theodore Roosevelt, Typical American: His Life and Work: Patriot, Orator, Historian, Sportsman, Soldier, Statesman and President.* L. H. Walter, 1919, 314.

Holder, J. B. "Address of Dr. J. B. Holder." In *The Third Annual Report of the Society of American Taxidermists, 1882–83.* Rochester, NY: Daily Democrat and Chronicle Book and Job Print, 1883.

Holland, W. J. *The Carnegie Museum, Annual Report of the Director for the Year Ending March 31, 1898.* Pittsburgh: Murdoch-Kerr Press, 1898.

———. *The Carnegie Museum, Annual Report of the Director for the Year Ending March 31, 1900.* Pittsburgh: Murdoch-Kerr Press, 1900.

Hornaday, William T. "The Crocodile in Florida." *American Naturalist* 9, no. 9 (September 1875): 498–504.

———. "Extermination of the American Bison." In *Annual Report of the Board of Regents of the Smithsonian Institution for the Year Ending June 30, 1887.* Washington, DC: Government Printing Office, 1889.

———. *Our Vanishing Wild Life.* New York: New York Zoological Society, 1912.

———. "The Passing of the Buffalo–I." *Cosmopolitan* 4 (October 1887): 91.

———. *Taxidermy and Zoological Collecting.* New York: C. Scribner's Sons, 1902.

———. *Thirty Years War for Wildlife.* New York: Charles Scribner's Sons for the Permanent Wildlife Protection Fund, 1931.

———. *Two Years in the Jungle.* New York: C. Scribner's Sons, 1886.

———. *A Wild Animal Round-Up.* New York: C. Scribner's Sons, 1925.

Horowitz, Helen Lefkowitz. "The National Zoological Park: 'City of Refuge' or Zoo?" In *New Worlds New Animals: From Menagerie to Zoological Park in the Nineteenth Century,* ed. Robert J. Hoage and William A. Deiss. Baltimore: Johns Hopkins University Press, 1996.

Hough, E. "Elk in Wisconsin." *Forest and Stream,* May 11, 1895, 369.

Huey, Laurence M. "Past and Present Status of the Northern Elephant Seal with a Note on the Guadalupe Fur Seal." *Journal of Mammalogy* 11, no. 2 (May 1930): 188–94.

Jordan, David Starr. *The Days of a Man: Being Memories of a Naturalist, Teacher and Minor Prophet of Democracy.* Vol. 1. New York: World Book, 1922.

"Kenneth E. Behring Hall of Mammals." In *National Museum of Natural History: Annual Report for 2003.* Available at http://www.mnh.si.edu.

Kenyon, Karl W., and Ford Wilke. "Migration of the Northern Fur Seal, *Callorhinus ursinus.*" *Journal of Mammalogy* 34, no. 1 (February 1953): 86–98.

Kes Hillman-Smith, A. K., and Colin P. Groves. "*Diceros bicornis.*" *Mammalian Species,* no. 455 (June 2, 1994): 1–8.

King, Geoff, Claire Malloy, and Yannis Tzioumakis. *American Independent Cinema: Indie, Indiewood and Beyond.* London: Routledge, 2013.

Kisling, Vernon M., Jr. "The Origin and Development of American Zoological Parks to 1899." In *New Worlds New Animals: From Menagerie to Zoological Park in the Nineteenth Century,* ed. Robert J. Hoage and William A. Deiss. Baltimore: Johns Hopkins University Press, 1996.

Kohler, Robert E. *All Creatures: Naturalists, Collectors, and Biodiversity, 1850–1950.* Princeton, NJ: Princeton University Press, 2006.

———. "Subspecies Classification and Biological Survey, 1850s–1930s." Max-Planck-Institut für Wissenschaftsgeschichte Preprint Series, no. 240 (2003).

Kohlstedt, Sally Gregory. "Henry A. Ward: The Merchant Naturalist and American Museum Development." *Journal of the Society for the Bibliography of Natural History* 9, no. 4 (1980).

Kohlstedt, Sally Gregory, and Paul Brinkman. "Framing Nature: The Formative Years of Natural History Museum Development in the United States." *Proceedings of the California Academy of Sciences* 55, no. 2, suppl. 1 (September 30, 2004): 7–33.

"The Last Night of a Successful Exhibition." *Rochester Democrat and Chronicle,* December 23, 1880. Reprinted in *Ward's Natural Science Bulletin* 1, no. 1 (June 1, 1881).

Leigh, William R. *Frontiers of Enchantment: An Artist's Adventures in Africa.* New York: Simon & Schuster, 1938.

———. "Painting the Backgrounds for the African Hall Groups." *Natural History* 30 (1930): 575.

Levinson, Judith, and Sari Uricheck. "Documenting the Documents: The Conservation Survey of the Akeley Hall of African Mammals." *Objects Specialty Group Postprints* 12 (2005): 39–61.

Lidicker, W. Z., Jr. "The Nature of the Subspecific Boundaries in a Desert Rodent and Its Implications for Subspecific Taxonomy." *Systematic Zoology* 11 (1962): 160–71.

Lucas, Frederic A. "Akeley as a Taxidermist." *Natural History* 27 (March–April 1927): 142–52.

———. "Animals Recently Extinct or Threatened with Extermination, as Represented in the Collections of the U.S. National Museum." In *Annual Report of the Board of Regents of the Smithsonian Institution for the Year Ending June 30, 1889.* Washington, DC: Government Printing Office, 1891.

———. "Breeding Habits of the Pribilof Fur Seal." In *The Fur Seals and Fur-Seal Islands of the North Pacific Ocean.* Washington, DC: Government Printing Office, 1899.

———. "The Expedition to Funk Island, with Observations upon the History and Anatomy of the Great Auk." In *Report of the U.S. National Museum for 1888.* Washington, DC: U.S. Government Printing Office, 1890.

———. *Fifty-Third Annual Report of the American Museum of Natural History for the Year 1921.* New York: The Museum, 1922.

———. *Fifty Years of Museum Work: Autobiography, Unpublished Papers, and Bibliography.* New York: American Museum of Natural History, 1933.

———. *Forty-Third Annual Report of the Trustees of the American Museum of Natural History for the Year 1911.* New York: The Museum, 1912.

———. "General Administration." In *Forty-Fourth Annual Report of the American Museum of Natural History for the Year 1912,* 30–38. New York: The Museum, 1913.

———. *Forty-Ninth Annual Report of the American Museum of Natural History for the Year 1917,* 45–50. New York: The Museum, 1918.

———. "The Fur Seal." *American Museum Journal* 12 (1912): 132–33.

———. "Glimpses of Early Museums." *Natural History,* January–February 1921.

———. "The Mounting of Mungo." *Science* 7, no. 193 (1886): 337–41.

———. *Museum News,* November 1905, 46.

———. *Museums of the Brooklyn Institute of Arts and Sciences: Report upon the Condition and Progress of the Museums for the Year Ending December 31, 1905.* New York: Brooklyn Institute, 1906.

———. "Official Extermination." *Forest and Stream* 28 (March 3, 1887): 104.

———. "The Passing of the Whale." *Supplement to the Zoological Society Bulletin,* July 1908.

———. "The Question of Groups." *Museum News,* April 1909, 97–98.

———. "The Scope and Needs of Taxidermy." In *The Third Annual Report of the Society of American Taxidermists, 1882–83.* Rochester, NY: Daily Democrat and Chronicle Book and Job Print, 1883.

———. "The Story of Museum Groups." *American Museum Journal,* January 1914.

———. *The Story of Museum Groups.* Guide Leaflet Series, no. 53. New York: American Museum of Natural History, 1921.

Mayr, Ernst. *Animal Species and Evolution.* Cambridge, MA: Belknap Press of Harvard University Press, 1963.

McKinley, Edward H. *The Lure of Africa: American Interests in Tropical Africa, 1919–1939.* Indianapolis: Bobbs-Merrill, 1974.

McLean, Marshall. "Discussion and Correspondence: The Pribilof Fur Seal Herd." *Science* 35, no. 892 (February 2, 1912): 183–84.

Merriam, C. Hart. "The Biological Survey—Origin and Early Days—A Retrospect." *Survey* 16, no. 3 (March 1933): 4.

Mitman, Gregg. *Reel Nature: America's Romance with Wildlife on Film.* Cambridge, MA: Harvard University Press, 1999.

Molina, Miquel. "More Notes on the Verreaux Brothers." *Pula: Botswana Journal of African Studies* 16 (2002): 30–36.

Morse, E. S. "Notes." *American Naturalist,* April 1873, 250.

"Museum Notes." *Science,* May 29, 1903, 873–74.

"The National Museum Buffalo." *Forest and Stream* 28, no. 6 (March 1877): 3.

"The Newly Discovered Elephant Seals." *Forest and Stream*, April 29, 1911, 652.

"Notes and News." *The Auk* 22 (1905): 109.

Nyhart, Lynn K. *Modern Nature: The Rise of the Biological Perspective in Germany.* Chicago: University of Chicago Press, 2009.

O'Brian, Dan. *Great Plains Bison*. Lincoln: University of Nebraska Press, 2017.

Osborn, Henry Fairfield. *Fifty-Second Annual Report of the American Museum of Natural History for the Year 1920*. New York: The Museum, 1921.

———. *Fifty-Seventh Annual Report of the American Museum of Natural History for the Year 1925*. New York: The Museum, 1926.

———. Foreword to *Our Vanishing Wild Life*, by William T. Hornaday, vii–viii. New York: New York Zoological Society, 1912.

———. *Forty-Eighth Annual Report of the American Museum of Natural History for the Year 1916*. New York: The Museum, 1917.

———. *Forty-Fifth Annual Report of the American Museum of Natural History for the Year 1913*. New York: The Museum, 1914.

———. "The New York Zoological Park and Aquarium." *Science* 17, no. 424 (February 13, 1903): 265.

———. "Preservation of the Wild Animals of North America." Address before the Boone and Crockett Club. Private Printing for the Boone and Crockett Club of Washington, 1904.

———. "Preservation of the World's Animal Life." *American Museum Journal* 12 (1912): 124.

———. "Report of the President." In *Forty-Fourth Annual Report of the Trustees of the American Museum of Natural History for the Year 1912*. New York: Irving Press, 1913.

Osborn, Henry Fairfield, and H. E. Anthony. "The Close of the Age of Mammals." *Journal of Mammalogy* 3, no. 4 (November 1922): 219–37.

Osgood, Wilfred H., Edward A. Preble, and George H. Parker. *The Fur Seals and Other Life of the Pribilof Islands, Alaska, in 1914*. Bulletin of the Bureau of Fisheries, vol. 34. Washington, DC: Government Printing Office, 1914.

"Our Centennial Letters—No. 8." *Forest and Stream*, August 3, 1876, 423.

"Our Group of Ornithorhynchus." *Ward's Natural Science Bulletin* 2, no. 2 (April 1, 1883): 9.

"Our Rochester Letter." *Forest and Stream*, December 23, 1880, 409.

Parkes, Kenneth C. "In Memoriam: Walter Edmond Clyde Todd." *The Auk* 87, no. 4 (October 1987).

Peale, Charles Willson. *The Autobiography of Charles Willson Peale*. Edited by Stanley Hart. Vol. 5 of *The Selected Papers of Charles Willson Peale and His Family*, edited by Lillian Miller, Sidney Hart, David C. Ward, Lauren E. Brown, Sara C. Hale, and Leslie K. Reinhardt. New Haven, CT: Yale University Press for the National Portrait Gallery, Smithsonian Institution, 2000.

Petterson, Palle B. *Cameras into the Wild: A History of Early Wildlife and Expedition Filmmaking, 1895–1928*. Jefferson, NC: McFarland, 2011.

Proceedings of the American Association of Museums. Vol. 1, *Records of the Meeting Held at the Museum of the Carnegie Institute, June 4–6, 1907*. Pittsburgh: The Association, 1908.

Proceedings of the American Association of Museums. Vol. 2, *Records of the Third Annual Meeting Held at Chicago, Illinois, May 5–7, 1908.* Charleston, SC: The Association, 1908.

Proceedings of the American Association of Museums. Vol. 9, *Records of the Tenth Annual Meeting Held in San Francisco, July 6–9, 1915.* Charleston, SC: The Association, 1915.

Reiger, John F. *American Sportsmen and the Origins of Conservation.* 3rd ed. Corvallis: Oregon State University Press, 2001.

Report of the U.S. National Museum under the Direction of the Smithsonian Institution for the Year Ending June 30, 1886. Washington, DC: Government Printing Office, 1889.

Report of the U.S. National Museum under the Direction of the Smithsonian Institution for the Year Ending June 30, 1889. Washington, DC: Government Printing Office, 1891.

Report upon the Condition and Progress of the U.S. National Museum during the Year Ending June 30, 1888. Washington, DC: Government Printing Office, 1889.

"Revolution in Taxidermy." *New York Commercial Advertiser,* May 3, 1883, reprinted in *Ward's Natural Science Bulletin* 2, no. 2 (April 1, 1883): 16.

Robins, Louise E. *Elephant Slaves and Pampered Parrots: Exotic Animals in Eighteenth-Century Paris.* Baltimore: Johns Hopkins University Press, 2002.

Rockwell, Robert H. *My Way of Becoming a Hunter.* New York: Norton, 1955.

Roosevelt, Theodore. *African Game Trails: An Account of the African Wanderings of an American Hunter-Naturalist.* New York: Charles Scribner's Sons, 1910.

Roper, Laura Wood. *FLO: A Biography of Frederick Law Olmsted.* 1973; repr., Baltimore: Johns Hopkins University Press, 1983.

Sacco, Janis C., and Duane A. Schlitter. "The Return of the Arab Courier: 19th-Century Drama in the North African Desert." *Carnegie Magazine* 62, no. 2 (1994): 31–32, 38–41.

"Scientific Journals and Articles." *Science,* November 10, 1905, 596.

"Scientific Notes and News." *Science,* December 31, 1897, 991.

"Scientific Notes and News." *Science,* September 3, 1909, 305.

The Second Annual Report of the Society of American Taxidermists, 1881–82. Rochester, NY: Daily Democrat and Chronicle Book and Job Print, 1882.

Sherwood, George H. *Fifty-Eighth Annual Report of the American Museum of Natural History for the Year 1926.* New York: The Museum, 1927.

———. *General Guide to the Exhibition Halls of the American Museum of Natural History.* Guide Leaflet Series of the American Museum of Natural History, no. 35. New York: American Museum of Natural History, 1911.

Shufeldt, R. W. "Scientific Taxidermy for Museums." In *Report of the National Museum for the Year Ending June 30, 1892.* Washington, DC: Government Printing Office, 1893.

"Sixteenth Congress of the American Ornithologists' Union." *The Auk,* January 1899, 54.

"Snap Shots." *Journal of Outdoor Life,* January 26, 1893, 1.

"The Society of American Taxidermists." *Scientific American* 48, no. 20 (May 19, 1883): 305.

"The Society of Taxidermists." *Ward's Natural Science Bulletin* 2, no. 1 (January 1, 1883): 2.

Star, Susan Leigh. "Craft vs. Commodity, Mess vs. Transcendence: How the Right Tool Became the Wrong One in the Case of Taxidermy and Natural History." In *The Right Tools for the Job: At Work in Twentieth-Century Life Sciences*, ed. Adele Clarke and Joan H. Fujimura. Princeton, NJ: Princeton University Press, 1992.

Stewart, Brent S., and Harriet R. Huber. "*Mirounga angustirostris*." *Mammalian Species* no. 449 (November 15, 1993): 1–10.

"A Story of Destruction." *Forest and Stream*, 31, no. 10 (1888): 181.

Supplement to the *American Journal of International Law*. Vol. 6. New York: Baker, Voorhis and Company for the Society of International Law, 1912.

"The Taxidermists' Exhibition." *Ward's Natural Science Bulletin* 2, no. 2 (April 1, 1883): 13.

"Taxidermy at Home." *Ward's Natural Science Bulletin*, January 1, 1883.

The Third Annual Report of the Society of American Taxidermists, 1882–83. Rochester, NY: Daily Democrat and Chronicle Book and Job Print, 1883.

Townsend, Charles H. "The California Sea-Elephant." *Forest and Stream*, January 13, 1887, 485.

———. "Discussion and Correspondence: The Pribilof Fur Seal Herd." *Science* 35, no. 896 (March 1, 1912): 334–36.

———. "The Elephant Seal Not Extinct." *Century*, June 1912, 205–11.

———. "Endurance of the Porpoise in Captivity." *Science* 43, no. 1111 (1916): 534–35.

———. "The Fate of the Whale." *Forest and Stream*, August 8, 1908, 2.

———. "In Memoriam: Frederic Augustus Lucas." *The Auk* 47 (April 1930): 147–58.

———. "Old Times with the Birds: Autobiographical with Two Portraits." *The Condor* 29, no. 5 (September–October 1927): 224–32.

———. "The Passing of the Whale": Editor's note. *Supplement to the Zoological Society Bulletin*, July 1908, 446.

———. "The Pribilof Fur Seal Herd and the Prospects for Its Increase." *Science* 34, no. 878 (October 27, 1911): 569.

———. "Sea Elephants on Exhibition." *Forest and Stream*, March 18, 1911, 412.

———. "Voyage of the 'Albatross' to the Gulf of California in 1911." *Bulletin of the American Museum of Natural History* 35 (1916).

———. "West Indian Seals at New York Aquarium." *Forest and Stream*, September 4, 1909, 372.

Townsend, Charles H., and George A. Clark. "Discussion and Correspondence: The Pribilof Fur Seal Herd." *Science* 35, no. 896 (March 1, 1912): 334–38.

True, Frederick W. "The Exhibition of Cetaceans by Papier Maché Casts." *Science* 8, no. 186 (July 22, 1898): 109.

Turner, Myles. *My Serengeti Years: The Memoirs of an African Game Warden*. New York: W. W. Norton, 1988.

Wallace, Joseph. *A Gathering of Wonders: Behind the Scenes at the American Museum of Natural History*. New York: St. Martin's Press, 2000.

Ward, Roswell Howell. *Henry A. Ward: Museum Builder to America*. Rochester, NY: Rochester Historical Society, 1948.

Webster, Frederic S. "The Birth of Habitat Groups: Reminiscences Written in His Ninety-Fifth Year." *Annals of the Carnegie Museum* 30 (1945).

————. "Taxidermy as a Decorative Art." In *The Third Annual Report of the Society of American Taxidermists, 1882–83*. Rochester, NY: Daily Democrat and Chronicle Book and Job Print, 1883.

Wheeler, William Morton. "Carl Akeley's Early Work and Environment." *Natural History* 27, no. 2 (March–April 1927): 133–41.

————. *Seventh Annual Report of the Board of Trustees of the Public Museum of the City of Milwaukee, October 1, 1889*. Milwaukee: Milwaukee Public Museum, 1890.

Williams, Stephen L., and Catharine A. Hawks. "History of the Preparation Materials Used for Recent Mammal Specimens." In *Mammal Collection Management*, ed. Hugh H. Genoways, Clyde Jones, and Olga L. Rossolimo, 21–49. Lubbock: Texas Tech University Press, 1987.

Winsor, Mary Pickard. "Agassiz's Notions of a Museum: The Vision and the Myth." In *Cultures and Institutions of Natural History: Essays in the History and Philosophy of Science*, edited by Michael T. Ghiselin and Alan E. Leviton, 249–71. San Francisco: California Academy of Sciences, 2000.

Wonders, Karen. *Habitat Dioramas: Illusions of Wilderness in Museums of Natural History*. Stockholm, Sweden: Almqvist & Wiksell, 1993.

"The World's Exposition: From a Sportsman's Standpoint." *Forest and Stream* 24, no. 4 (February 19, 1885): 64.

"Zoological Society Annual Meeting." *Forest and Stream*, January 20, 1912, 83.

INDEX